A^LARMIN^G!

THE CHASM SEPARATING
EDUCATION OF APPLICATIONS
OF FINITE MATH FROM ITS
NECESSITIES

A^lARMING!
THE CHASM SEPARATING EDUCATION OF APPLICATIONS OF FINITE MATH FROM ITS NECESSITIES

Beware the Assumptions,
It's All in the Assumptions

WILLIAM J. ADAMS
Mathematics Department, Pace University

with illustrations by
Ramunė B. Adams

Library of Congress Control Number: 2013903350
ISBN: Hardcover   978-1-4797-9993-0
                    Softcover   978-1-4797-9992-3
                    Ebook   978-1-4797-9994-7

This book was printed in the United States of America.

To order additional copies of this book, contact:
Xlibris LLC
1-888-795-4274
www.Xlibris.com
Orders@Xlibris.com
116073

To Aleksa, Gaja, Sabrina, and Veronika

Acknowledgment: I am indebted to my daughter Ramunė for preparing the illustrations.

PREFACE

As an introduction to applied mathematics the sample of recently published finite math books that I gave thought to have two major shortcomings that make them unsatisfactory for this purpose.

1. Mathematical modeling which distinguishes a conclusion's validity from its realism is not taken up.

2. The issue, if the hypothesis of a theorem in an application setting is not satisfied, what then?, is not addressed.

Truth in book title description requires that the afore sample of finite math books that I examined be called finite math with **illustrations** rather than **applications** since the difference is significant.

The purpose of this book is to address the afore dimensions of applied finite mathematics by providing illustrations and food-for-thought questions with answers/discussion to assist colleagues and students who share my concerns to become better acquainted with these dimensions.

As a point of information, for discussion of these dimensions in a finite math book I note W.J. Adams, *Finite Mathematics, Models, and Structure*, Revised Edition (Xlibris, 2009).

W.J.A.

BRIEF CONTENTS

CONTENTS

PART 1

Is Math Modeling a Necessity?

PREFACE

YES, absolutely. Consider the difference between It's a Word Problem View and It's a Math Model View.

Word Problem View

> 1. No distinction is made between the mathematical validity and real-world realism of the solution obtained.
>
> 2. The underlying statements that lead to valid conclusion are viewed as (real-world) facts, not as **assumptions** whose real-world realism may be open to question.
>
> 3. Since a valid conclusion viewed as based on facts is a fact, there is no question about implementing it.

Math Model View

1. Mathematical modeling rests on a distinction made between the mathematical validity of a conclusion and its real-world realism.

2. What is viewed as facts from the word problem view is viewed as **assumptions** whose realism may be open to question.

   If the **assumptions** are unrealistic, the realism of a valid conclusion is open to question.

3. If the **assumptions** are realistic, this provides a GO-AHEAD to implementing a valid conclusion.

   If some of the **assumptions** are not realistic, this provides a RED-ALERT that the realism of any valid conclusion is open to question and that its implementation might prove to be disastrous.

Shades of Validity: Be on Guard

I employ the term **valid** in this book in a very specific manner—roughly, that the so labeled statement follows as an inescapable consequence of a number of statements taken as a starting point (called a **hypothesis**), and individually termed **assumptions**, **postulates** or **axioms**. It is this sense of the word valid that is to be understood throughout the book.

As is the case with many words, valid is used in a number of ways in our everyday language. We might hear, for example, Jim had valid reasons for not following instructions; we must validate our assumptions. Uses for the word valid in our language include sound, cogent, convincing, telling and, unfortunately for discussions of applications of mathematics, true/realistic.

In reading any exposition some degree of misunderstanding is to be expected because of the many ways that words are used in our language. Beware.

1

Is It Just a Word Problem, or Is There More to It Than That?

1.1 PREFACE

In the 1960's mathematics and mathematics education organizations endorsed mathematical modeling as part of their New Math proposals. The acceptance and implementation of this part of their propsals was very limited. The focus of mathematics education continued to involve translating a problem and its background into math form, solving it, and noting how to implement the solution. My book *Elements of Linear Programming*[1] is a book of this kind. A review[2] noted that "The authors present an elementary account of linear programming and two-person zero-sum games with the primary intent being to teach the non-mathematics major how to set up and solve linear programs. . . . this reviewer feels that the authors have selected an interesting topic and presented it on a level which non-mathematics majors should be able to understand."

It may not seem surprising that what was standard mathematics education half a century ago will not suffice for the needs of the 21st century. To obtain a concrete sense in considering applications of finite mathematics I invite you to consider the case of the Last National Bank.

1.2 THE LAST NATIONAL BANK

The Last National Bank has assets in the form of loans and securities that, it is **assumed**, bring returns of 10 and 8 percent, respectively, in a certain

---

[1]   Coauthored with Allan Gewirtz and Louis V. Quintas (Van Nostrand Reinhold Co., 1969), first appeared in preliminary and preprint editions.

[2]   *Journal of the Mathematics Association of Two-Year College Teachers*, Spring 1971

time period. The bank has a total of $60 million to allocate between loans and securities. Two major guidelines are imposed by the bank on its lending activity: (1) a securities balance equal to or greater than 25 percent of total assets must be maintained, and (2) at least $15 million must be available for loans.

The bank is interested in determining, with respect to these conditions, how funds should be allocated between loans and securities so that investment income is maximized.

It's Just a Word Problem Point of View

Set up the problem as a linear program, solve it, and state what must be done to implement the solution, with what result.

Let x and y denote the amounts (in millions of dollars) to be invested in loans and securities, respectively. We have the following linear program:

$$\text{Maximize } I(x, y) = 0.10x + 0.08y$$
$$\text{subject to}$$
$$x \geq 0, y \geq 0$$
$$x + y \leq 60$$
$$-x + 3y \geq 0$$
$$x \quad\quad \geq 15$$

The solution, obtained by means of the corner point theorem, is $x = 45$, $y = 15$. The maximum value is 5.7. To implement this solution invest 45 million dollars in loans and 15 million dollars in securities. The expected return is 5.7 million dollars, and this is the maximum return that can be expected.

Math Modeling Point of View

Set up the problem in terms of a linear program model. What **assumptions** underlie the model? Solve the linear program of the model. Should the solution be implemented?

Let x and y denote the amounts (in millions of dollars) to be invested in loans and securities, respectively.

The **assumptions** that loans and securities bring returns of 10 and 8 percent, respectively, in the time period being considered, together with the conditions stated yield the linear program model with linear program

$$\text{Maximize } I(x, y) = 0.10x + 0.08y$$
subject to
$$x \geq 0, y \geq 0$$
$$x + y \leq 60$$
$$-x + 3y \geq 0$$
$$x \qquad \geq 15,$$

whose solution is (45,15) with maximum value 5.7.

Should this solution be implemented? William Ganz, vice president for investment allocation of the Last National Bank, asked for opinions.

1.   "Yes," said Jim Turner, the resident computer "expert". "The solution was obtained by use of the latest computer technology available, and that's good enough for me."

      Would you agree with Jim?, Bill Ganz was asked. No, he replied. The latest computer technology will give us that the valid conclusion (45,15) is the solution and 5.7 is the maximum value. But its use does not address the issue of the realism of the **assumptions**, which is our concern.

2.   "Of course, no question about it," said June Carver, chief financial officer. "The corner point method is the mathematical basis for this solution and its use ensures that profit will be maximized."

      Would you agree with June? Bill, was asked. No, he replied. Same answer as the afore. Replace latest computer technology by corner point method.

3.   "Are there any factors that we should have taken into consideration in the formulation of this linear program model," Ganz asked Horace Black, head of the quantitative analysis department.

> Yes, replied Horace. The realism of the **assumptions** that loans and securities will bring returns of 10 and 8 percent, respectively, in the time period under consideration. Based on the analysis we carried out we concluded that these **assumptions** are not realistic.

> The solution is a valid consequence of the model, mathematically speaking, that's not at issue. But since it is based on unrealistic **assumptions** it too might be unrealistic, and to implement it might prove disastrous to the Bank.

> My team and I recommend that you do not implement this solution and have us work on refining the model to make it realistic.

Needless-to-say, Horace's judgment prevailed.

1.3   A FEW OBSERVATIONS

Math Modeling is a more realistic approach than It's-Just-a-Word-Problem to the application dimension of finite math (and all math, for that matter), but it places a heavy burden on what Agatha Christie's Hercule Poirot termed "the little grey cells".

From my experience most students find it difficult to adopt to this thinking. Their experience is that "math is precise" in the sense that one question, one answer, and that's that, which is why they feel comfortable with the It's-Just-a-Word-Problem approach. To inform them that in the application dimension of finite math there is more to it than one question, one answer, no ifs-ands-or-buts, is counter to their math studies.

For my colleagues and I it's a challenge to ease them into this very different dimension of applied math. From my experience the task requires time, patience, a variety of simple to understand examples, and what I term food-for-thought questions that pose a challenge to "the little grey cells".

2

Which, if Either, Is the "Right" Linear Program Model?

2.1 PREFACE

What is the meaning of "Right" in this chapter's title? The Austin company's production scheduling problem is used as a vehicle to address this question.

2.2 THE AUSTIN COMPANY'S PROBLEM

The Austin Company, a producer of high quality electronic home entertainment equipment, has decided to enter the digital tape player market by introducing two models, DT-1 and DT-2, to be called ultra and supreme when the marketing campaign is put into operation. Their problem is to determine the number of units of each model to be produced to maximize profit.

To follow up on the math modeling framework in this setting two linear program models are introduced.

The Operations Research Department's Model, LP-1

The Company's operations research department was asked to study the situation and make recommendations. The OR department began their analysis by collecting data. They divided the manufacturing process into three phases; construction, assembly, and finishing. The data collected and their analysis led them to introduce the following **assumptions/postulates**.

> P1. In the construction phase each DT-1 unit requires 2 hours of labor and each DT-2 unit requires 3 hours of labor. At most 1,100 hours of construction time are available per week.

P2. In the assembly phase each DT-1 unit requires 5 hours of labor and each DT-2 unit requires 3 hours of labor. At most 1,400 hours of assembly time are available per week.

P3. In the finishing phase each DT-1 unit requires 4 hours of labor and each DT-2 unit requires 1 hour of labor. At most 756 hours of finishing time are available per week.

P4. After taking cost and revenue factors into consideration the anticipated profit for each DT-1 unit is $150 and the anticipated profit for each DT-2 unit is $120. In order for these unit profit values to be realistic the Company must produce at least 25 DT-1 and 40 DT-2 units per week.

P5. There is an unlimited market for the DT-1 and DT-2 models.

P. All factors other than the ones considered in the analysis of the production of the DT-1 and DT-2 models are negligible.

Its next task was to translate these postulates into mathematical form, being careful to include everything stated in the postulates and not go beyond them. The OR department began by introducing variables for the quantities it sought to determine; let x denote the number of DT-1 and y the number of DT-2 units to be made weekly. There is a fair amount of data contained in the postulates and it is useful to make it available at a glance in tabular form. This is done in Table 2.1.

Table 2.1

	Nu of units to be made per week	Profit per unit	Construction time per unit	Assembly time per unit (hrs)	Finishing time per unit (hrs)
DT-1	x	$150	2	5	4
DT-2	y	$120	3	3	1
			≤ 1100	≤ 1400	≤ 756

$$profit = \begin{bmatrix} profit\ on \\ DT-1 \end{bmatrix} + \begin{bmatrix} profit\ on \\ DT-2 \end{bmatrix}$$

$$= \begin{bmatrix} profit\ on \\ one\ DT-1 \\ unit \end{bmatrix} \cdot \begin{bmatrix} no.\ of \\ units \\ made \end{bmatrix} + \begin{bmatrix} profit\ on \\ one\ DT-2 \\ unit \end{bmatrix} \cdot \begin{bmatrix} no.\ of \\ units \\ made \end{bmatrix}$$

$$= 150x + 120y$$

The profit obtained by making x DT-1 and y DT-2 units per week is expressed by the linear function:

$$P(x,\ y) = 150x + 120y$$

As to the conditions that x and y must satisfy, since the number of units made must be non-negative, we have:

$$x \geq 0$$
$$y \geq 0$$

The construction time condition is that

$$(\text{total construction time used}) \leq 1100.$$

In terms of unit construction times, 2 hours are needed for one unit of DT-1 and 3 hours are needed for one unit of DT-2, $2x$ hours are needed for x DT-1 units and $3y$ hours are needed for y DT-2 units. The total construction time utilized is expressed by $2x + 3y$. Therefore, the construction time utilized is expressed by $2x + 3y$. Therefore, the construction time condition is:

$$2x + 3y \leq 1100$$

Similarly, the assembly and finishing time conditions are stated by the inequalities:

$$5x + 3y \leq 1400$$
$$4x + y \leq 756$$

The conditions that at least 25 DT-1 and 40 DT-2 units must be produced weekly are expressed by the inequalities:

$$x \geq 25$$
$$y \geq 40$$

We emerge with the following mathematical structure, call it **linear program model LP-1**, as a translation of the **postulates** introduced by the OR department of the Austin Company.

$$\text{Maximize } P(x, y) = 150x + 150y$$
subject to
$$x \geq 0, y \geq 0$$
$$2x + 3y \leq 1100$$
$$5x + 3y \leq 1400$$
$$4x + y \leq /56$$
$$x \geq 25, y \geq 40$$

Here x represents the number of DT-1 and y the number of DT-2 units to be made weekly.

A few definitions and an observation are useful:

A **linear program** is a mathematical problem with the following structure: there is specified a **linear function** of a number of variables that are required to satisfy linear conditions described by some mixture of **linear inequalities** and **linear equations**, called **constraints**.

The problem is to find values for these variables which satisfy the constraints, called **feasible points** and yield the maximum, or minimum, value of function, which is called an **objective function**.

A linear program may or may not arise from a real world situation/ problem under study. If it does, the linear program together with the **assumptions/postulates** that led to it is called a **linear program model** for the situation/problem under study.

The Aleksa Company's Model, LP-2

The Austin Company also hired the Aleksa Company, a consulting operations research firm, to independently study the digital tape player situation and make recommendations. The Aleksa OR group divided the manufacturing process into two phases: construction (which included assembly) and finishing. The data collected and their analysis led them to introduce the following **postulates**.

P1a. In the construction phase each DT-1 unit requires 8 hours of labor and each DT-2 unit requires 5 hours of labor. At most 2,210 hours of construction time are available per week.

P2a. In the finishing phase each DT-1 unit requires 3 hours of labor and each DT-2 unit requires 2 hours of labor. At most 860 hours of finishing time are available per week.

P3a. The anticipated profit for each DT-1 unit is $140 and the anticipated profit for each DT-2 unit is $150. In order for these unit profit values to be realistic the company must produce at least 50 DT-1 and 50 DT-2 units per week.

P4a. There is an unlimited market for the DT-1 and DT-2 models.

P.   All factors other than the ones considered in the analysis of the production of the DT-1 and DT-2 models are negligible.

The same sort of analysis that leads to LP-1 from the **postulates** introduced by the Austin Company's operation research department leads to the Aleksa OR group's **linear program model LP-2**:

$$\text{Maximize } P(x, y) = 140x + 150y$$
subject to
$$x \geq 0, y \geq \quad 0$$
$$8x + 5y \leq 2210$$
$$3x + 2y \leq \quad 860$$
$$x \geq 50, y \geq \quad 50$$

where x represents the number of DT-1 and y the number of DT-2 units to be made weekly.

2.3   SOLUTIONS OF MODELS LP-1 AND LP-2

Solution of LP-1

Our first step is to sketch the graph of the feasible points.

Locate the corner points on the graph and solve the appropriate systems of equations to determine their coordinates. There are five corner points, shown in Figure 2.1: (25, 40), (25, 350), (100, 300)—obtained by solving $2x + 3y = 1100$ and $5x + 3y = 1400$, (124, 260)—obtained by solving $4x + y = 756$ and $5x + 3y = 1400$, and (179, 40).

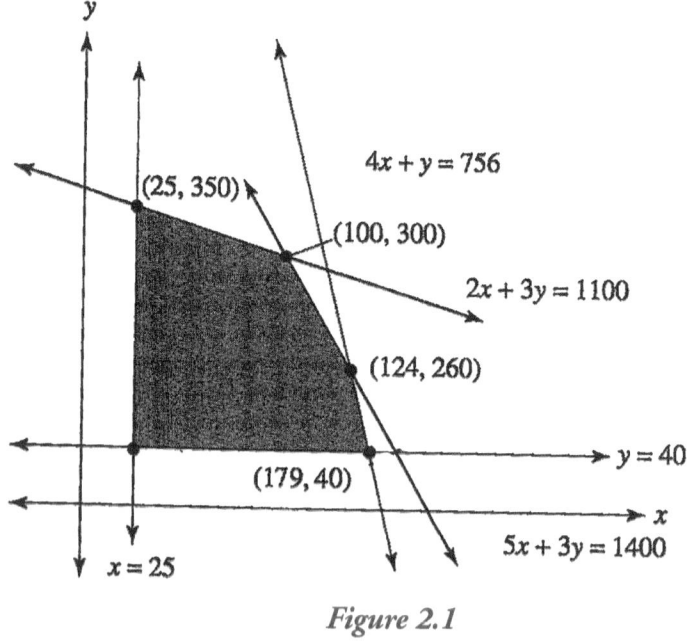

Figure 2.1

The computation of $P(x, y) = 150x + 120y$ at the corner points, summarized in Table 2.2, yields the solution (100, 300) with maximum value 51,000.

Table 2.2

Corner Point	$P(x, y) = 150x + 120y$
(25, 40)	8,550
(25, 350)	45,750
(100, 300)	51,000
(124, 260)	49,800
(179, 40)	31,650

Solution of LP-2

Locate the corner points on the graph of the feasible points, and solve the appropriate systems of equations to determine coordinates. There are four corner points, shown in Figure 2.2; (50, 50), (245, 50), (120, 250), and (50, 355).

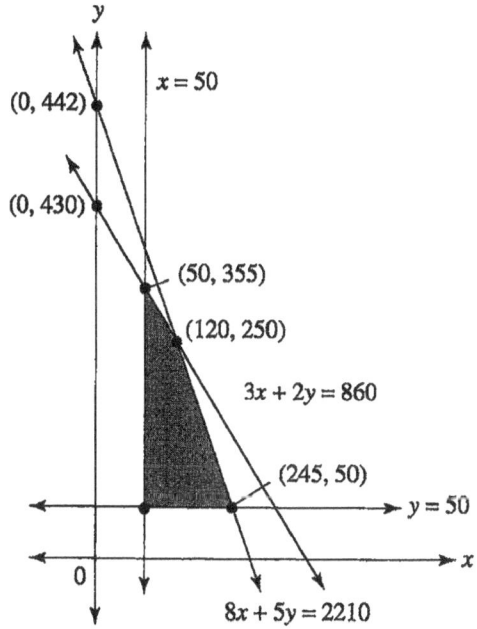

Figure 2.2

The computation of $P(x, y) = 140x + 150y$ at the corner points, summarized in Table 2.3 yields the solution (50, 355) with maximum value 60, 250.

Table **2.3**

Corner Point	$P(x, y) = 140x + 150y$
(50, 50)	14,500
(245, 50)	41,800
(120, 250)	54,300
(50, 355)	60,250

2.4 WHICH SOLUTION, IF EITHER, SHOULD BE IMPLEMENTED?

Model LP-1 has solution (100, 300) with maximum value 51,000 whereas model LP-2 has solution (50, 355) with maximum value 60,250. In terms of the Austin Company's situation, to implement LP-1's conclusion the production schedule would have to be set to manufacture 100 DT-1 and 300 DT-2 units per week with an anticipated weekly profit of $51,000; to implement LP-2's conclusion the production schedule would have to be set to manufacture 50 DT-1 and 355 DT-2 units per week with an anticipated weekly profit of $60,250.

Which solution, if either, should be implemented?

Bottom-line Bob vs. Reflective Ramunė

Bottom-line Bob is a product of the It's-Just-a-Word-Problem school of thought whereas Reflective Ramunė was educated in the Math-Modeling approach.

It's-Just-a-Word-Problem Recommendation

Bottom-line Bob, chairman of the ten member board charged with making a decision on how to implement the Company's entry into the digital tape player market, argued that it's obvious what we should do. "Implementation of LP-2 brings us a weekly profit of $60,250, whereas implementation of LP-1 brings us a weekly profit of $51,000. Since we want the largest possible return, we should go with LP-2. You can't make it any simpler." The board voted nine to one to implement LP-2.

Alas, the $60,250 weekly profit was far from being realized after LP-2 was implemented, and two years later the Austin Company's venture into the digital tape player market had to be written off as a disaster.

Math-Modeling Point of View

Bottom-line Bob, who presided over this disaster was confused, upset, and out of a job. He went to Reflective Ramunė, the chair of the new board and the one person who had voted against implementation of LP-2, with some questions: "We were ultra-cautious and obtained the additional services of the Aleksa OR group to make recommendations which, subsequently, had disastrous consequences for us; what went wrong? How is it that mathematics failed us?

Why did you vote against implementation of LP-2?" "Well Bob, as I pointed out at the board meeting, I voted against implementation of LP-2 because I was not convinced that its promise of a \$60,250 weekly profit was realistic. As you yourself pointed out, 'the promise of LP-2 is \$9,250 better than that promised by LP-1,' but promises may not be realizable if they are founded on unrealistic **assumptions**. The conclusion reached from LP-2 was indeed tempting, and in fact proved too tempting for my colleagues on the board, but since it came from a linear program model founded on **assumptions** which I viewed as unrealistic, I resisted temptation. We have no quarrel with mathematics; mathematics gave us a valid conclusion from LP-2, which is all that we can legitimately expect. Unfortunately that conclusion proved to be unrealistic."

2.5   Is Mathematics Precise?

"Ramunė, I don't understand this. I always liked math in high school and college. Solving those equations, factoring those expressions, differentiating those functions, throwing the data into the computer and letting it do its thing, that was real fun. What I like most about math is its precision. You don't get ten sides to a story. You get one answer and that's that; no baloney".

"Bob, I think your math

courses may have focused too much on technique and not enough on perspective. Technique can be fun to a point, but without a perspective on its place in the over-all role of mathematics in applications, we see only a small tip of the mathematical iceberg.

Mathematics is precise in the sense that it gives us valid conclusions based on the **assumptions** made, which is where technique—factoring, differentiating functions, and the like—plays its major role. Whether the **assumptions** made are realistic or not is another matter which technique can't help us with. The question of how to formulate these **assumptions** and reach a judgment on their realism may indeed yield ten sides to the story. I'm afraid that those who find mathematics attractive because of what they perceive to be its absolutist nature have misunderstood the meaning of mathematical precision."

2.6 HOW COULD IT BE WRONG? I USED A COMPUTER

"Ramunė, I still don't fully understand what went wrong. The company just spent millions to update its computer system. I had access to the latest and the best. Why didn't this save us from disaster."

"Bob, Henry Clay's observation that 'statistics are no substitute for judgment' applies equally well to the computer.

We cannot expect the computer to employ technological alchemy and convert unrealistic **assumptions** into golden truths. Keep in mind the GIGO principle; if garbage in, then garbage out. Indiscriminate use of computer technology has made possible the generation of more nonsense more quickly than ever before by more people having less understanding of what they are doing."

The Computer's Right of Way

"If what you say is true Ramunė, then what good is this super computer technology to us?" "For number crunching and delivering results quickly and efficiently, the computer is without equal, Bob.

In this dimension it is the undisputed master of the field. The mathematical model building process and computers have developed a symbiotic relationship in that computers have made it possible for us to solve previously unapproachable large scale problems that come out of mathematical models, while the accessibility of such

problems to computer solution has made possible the use of such complex models. Alas, none of this overrides the GIGO principle."

2.7   FOOD FOR THOUGHT

1.   **ZKB Electronics**

ZKB Electronics makes two kinds of personal computers, model ZKB-47 and model ZKB-82. The management of ZKB called in the Rex Consulting Firm to determine how many units of each model should be made daily to maximize profit. The consulting firm set up a linear program model for the electronics company's production problem and, by applying the corner-point method, reached the conclusion that 300 ZKB-47 units and 250 ZKB-82 units should be made daily to maximize profit. Before implementing this conclusion, the management of ZKB put the following questions to the director of the consulting firm.

(a)   Does use of the corner-point method guarantee that profit will be maximized when 300 ZKB-47 and 250 ZKB-82 units are made daily and sold?

(b)   What is your basis for recommending that we implement your conclusion?

How would you reply to these questions?

2.   **The Onutė Corporation's Production Problem**

The Onutė Corporation plans to introduce two high resolution TV models, T20 and T24, to the market. Its own operations research group was led to introduce the following M1 model to determine the optimal production schedule for maximizing profit:

$$\text{Maximize } P(x, y) = 180x + 120y$$
subject to
$$x \geq 0, y \geq 0$$
$$4x + 3y \leq 320$$
$$5x + 2y \leq 330,$$

where, x and y denote the number of T20 and T24 units, respectively, to be made daily. Its solution is (50, 40) with maximum value 13,800

The Aleksa Company was also hired to study the Onutė Corporation's production scheduling problem. It was led to introduce the following M2 model to determine the optimal production schedule for maximizing profit:

$$\text{Maximize } P(x, y) = 190x+110y$$
subject to
$$x \geq 0, \, y \geq 0$$
$$5x + 2y \leq 330$$
$$3.25x + 2y \leq 225$$
$$4x + 3y \leq 320,$$

where x and y denote the number of T20 and T24 units, respectively, to be made daily. Its solution is (60, 15) with maximum value 13,050

The following questions have arisen. How would you answer them?

(a) If mathematics is the precise subject that it is reputed to be, should there not be one solution to this problem rather than two?

(b) Since two solutions emerge, does it follow that not both are valid?

(c) Before making a decision about whether to implement M1 or M2, what questions would you put to the two operations research groups?

(d) Which model, if either, would you adopt and implement? Why? Is it possible that you would not adopt either model?

(e) The management of the Onutė Corp., educated in the It's-Just-a-Word-Problem school of thought implemented

M1 with a larger anticipated daily profit than M2. Disastrous financial consequences followed and their venture into the introduction of TV models T20 and T24 had to be abandoned.

If they came to you for an explanation of what went wrong, what would you tell them?

3.   **Environmental Protection**

The Saxon Company must produce at least 250 thousand tons of paper annually. From the current operating system 10 pounds of chemical residue is deposited into a neighboring water system for each ton of paper produced. The resulting pollution has become a problem of serious concern, and to remain eligible for state tax benefits the Saxon Company must restrict the chemical residue emitted into the state's water system to not exceed 200 tons per year. Two filtration systems, Delta and Beta, have emerged for consideration. It is estimated that the installation of the Delta system would reduce emissions to 2 pounds for each ton of paper produced, and installation of the Beta system would reduce emissions to 1 pound for each ton of paper produced. Capital and operating costs for the Delta and Beta systems have been estimated at $8 and $12, respectively, per ton of paper produced.

Problems, Questions

(a)   The problem is to determine how many tons or paper should be produced subject to the Delta system and how many should be produced subject to the Beta system so that the emissions standard is met at minimal cost.

(b)   What are the underlying **assumptions/postulates?**

(c)   Set up a linear program model for this problem based on its underlying **assumptions/postulates:**

(d)   Determine the solution and minimum value of the afore L.P. model.

(e)   Should this solution be implemented?

(f)   If YES to (e), how would this solution be implemented?

(g)   If NO to (e), what then?

4.   The Charles National Bank

The Charles National Bank has assets in the form of loans and negotiable securities which, it is **assumed**, bring returns of 10 and 8 percent, respectively, in a certain time period. The bank has a total of $60 million to allocate between loans and securities. To meet unanticipated deposit withdrawals the bank maintains a securities balance greater than or equal to 25 percent of total assets. Lending is the bank's most important activity and to satisfy its clients it requires that at least $15 million be available for loans.

Problems, Questions

(a) The Bank wishes to determine how funds should be allocated to maximize total investment income.

(b) What are the underlying **assumptions/postulates**?

(c) Set up a linear program model for this problem based on its underlying **assumptions/postulates**.

(d) Determine the solution and maximum value of the afore L.P. model.

(e) Should this solution be implemented?

(f) If YES to (e), how would this solution be implemented?

(g) If NO to (e), what then?

5. **How Many Coats and Dresses Should Be Made?**

The Hoffman Clothing Manufacturers, Inc., has available 120 square yards of cotton and 100 square yards of wool for the manufacture of coats and dresses. Two square yards of cotton and 4 square yards of wool are used in making a coat while 4 square yards of cotton are used in making a dress. Cotton costs $5 per square yard and wool cost $20 per square yard. Four hours of labor are needed to make a coat and 2 hours of labor are needed to make a dress. The cost of labor is $25 per hour. At most 110 hours of labor are available for the manufacture of the coats and dresses. A coat sells for $300 and a dress sells for $140.

Problems, Questions

(a) The problem is to determine how many coats and dresses should be made to maximize net income.

(b) What are the underlying **assumptions/postulates**?

(c)　Set up a linear program model for this problem based on its underlying **assumptions/postulates**.

(d)　Determine the solution and maximum value of the afore L.P. model.

(e)　Should this solution be implemented?

(f)　If YES to (e), how would this solution be implemented?

(g)　If NO to (e), what then?

6.　Certitude?

Do you agree or disagree with the following point of view. So state and explain in appropriate detail.

> "Mathematical methods have the advantage of certitude. No qualified person can resist the truth of a mathematical conclusion properly communicated. The job of communication may be difficult if the solution is complex, but when communication is competent, agreement is inevitable. If anyone doubts a solution, he can recalculate the equations and check the steps in the derivation. Then he must either demonstrate that there has been an error or acknowledge the truth of the solution."

2.8　ANSWERS/DISCUSSION OF FOOD FOR THOUGHT

1.　ZKB Electronics

(a)　The corner-point method does not guarantee that profit will be maximized when 300 ZKB-47 and 250 ZKB-82 units are made daily and sold. Whether these output levels or other ones will maximize profit is a question of truth, and the corner-point method, as a mathematical technique, can only ensure the validity of the predicted output levels with respect to the linear program model that was set up for the profit maximization problem in question. If the **assumptions** that the model reflects are realistic, then the aforementioned output levels, obtained as a valid conclusion of these assumptions, will maximize profit; if these **assumptions** are not realistic, then it might well happen that other output levels would yield a higher profit than that projected for 300 ZKB-47 and 250 ZKB-82 units.

(b)   The basis for implementing the conclusion that 300 ZKB-47 and 250 ZKB-82 units be made and sold daily is the belief, based on an analysis of the company's operations and the market, that the **assumptions** made are realistic.

2.   The Onutė Corporation's Production Problem

(a)   and (b). Mathematics is precise in the sense that it yields valid conclusions with respect to an underlying starting point called a **hypothesis**. If there are **two underlying hypotheses**, then two quite different valid conclusions—solutions to linear program models in this case—might arise, each valid with respect to its own starting point.

(c)   What **assumptions** underlie M1 and M2? On what basis were they made? Convince me that they are realistic.

(d)   I would adopt that model based on what I judged to be realistic **assumptions**. If I were not convinced that the **assumptions** for either of M1 and M2 were realistic, I would not adopt either model. I would call for additional studies.

(e)   The issue which you did not give thought to is whether the **assumptions/postulates** underlying M1 are realistic. If not, which turned out to be the case, the valid conclusion arising from M1 may not be realistic. This was the case with the disastrous financial consequences that you experienced.

3.   **Environmental Protection**

(b)   The installation of Delta and Beta would reduce emissions to 2 pounds and 1 pound respectively, for each ton of paper produced.

Capital and operating costs for the Delta and Beta systems are $8 and $12, suspectively, for each ton of paper produced.

Factors other than those considered above are negligible to minimizing cost.

(c)   Let x and y denote the number of tons of paper to be produced annually subject to the Delta and Beta systems, respectively. The basic data and conditions are summarized in Table 2.4.

Table 2.4

	Nu. of tons produced	Emissions per ton (lbs)	Operating cost per ton ($)
Delta	x	2	8
Beta	y	1	12
	$\geq 250,000$	$\leq 400,000$	

The cost function to be minimized is

$$C(x, y) = 8x + 12y.$$

The condition that the Saxon Company must produce at least 250 thousand tons of paper annually is expressed by

$$x + y \geq 250,000.$$

The total amount of chemical residue produced annually is the number of pounds produced through use of the Delta system, $2x$ pounds, plus the amount produced through use of the Beta system, y pounds. Since this cannot exceed 200 tons (400,000 lbs), we have

$$2x + y \leq 400,000.$$

We emerge with the linear program:

$$\text{Minimize } C(x, y) = 8x+12y$$

subject to
$$x \geq 0, y \geq 0$$
$$x + y \geq 250,000$$
$$2x + y \leq 400,000,$$

which, together with the afore **assumptions/postulates**, is the linear program model.

(d)   Solution (150,000; 100,000); minimum value 2,400,000.

(e)   Yes, if the afore **assumptions/postulates** are determined to be realistic. No, if they are not.

(f)   Produce 150,000 tons of paper subject to the Delta filtration system and 100,000 tons subject to the Beta filtration system.

(g)   The **assumptions/postulates** would have to be reexamined and refined with the objective of obtaining realistic **assumptions/postulates**.

4.   The Charles National Bank

(b)   In the time period under consideration loans and negotiable securities (money market funds, for example) bring returns of 10 and 8 percent respectively.

(c)   Let x and y denote the amount, in millions of dollars, to be allocated for loans and securities, respectively. The income function to be maximized is

$$I(x, y) = 0.10x + 0.08y.$$

The following constraints emerge:

$x + y \leq 60$: $60 million is available for investment in loans and securities.

$y \geq (1/4)(x + y)$,   or equivalently, $-x + 3y \geq 0$: A securities balance greater than or equal to 25% of total assets must be maintained. Note, total assets is defined as the sum of the amounts invested in loans and securities, which is $x + y$.

$x \geq 15$:   at least $15 million must be available for loans.

We obtain the linear program model

$$\text{Maximize } I(x, y) = 0.10x + 0.08y$$
subject to
$$x \geq 0, y \geq 0$$
$$x + y \leq 60$$
$$-x + 3y \geq 0$$
$$x \geq 15,$$

along with the **assumptions/postulates** noted in (b).

(d) Solution (45, 15); maximum value 5.7.

(e) Yes, if the afore **assumptions/postulates** are found to be realistic, no, if they are not.

(f) Allocate $45 million to loans and $15 million to securities. The anticipated interest on investment is $5.7 million.

(g) The **assumptions/postulates** would have to be reviewed and refined with the objective of replacing them by realistic **assumptions/postulates**.

Suppose that this review led to the **assumptions/postulates** that loans and securities would bring returns of 9 and 11 percent, respectively. Then the refined model would consist of the linear program

$$\text{Max. } I(x, y) = 0.09x + 0.11y$$
subject to
$$\text{same constraints,}$$

together with the modified **assumptions/ postulates** about the interest rates.

Of Interest

A. Broaddus, "Linear programming: A New Approach to Bank Portfolio Management," *Federal Reserve Bank of Richmond: Monthly Review,* vol. 58, No. 11 (Nov. 1972), pp. 3-11. This article provides an introductory nontechnical discussion of linear programming for bank portfolio management.

K.J. Cohen and F.S. Hammer, "Linear Programming and Optimal Bank Asset Management Decisions," *Journal of Finance,* vol. 22 (May 1967), pp. 147-165. This paper describes a linear program model that had been used for several years by Bankers Trust Company in New York to assist in reaching portfolio decisions.

5.   How Many Coats and Dresses Should Be Made?

(b)   Cotton costs $5 per square yard and wool costs $ per square yard. The cost of labor is $25 per hour. At most 110 hours of labor are available. A coat sells for $300 and a dress sells for $140.

(c)   Let x and y denote the number of coats and dresses to be made, respectively. The data given are summarized in Table 2.5

Table 2.5

	Selling price	Cotton used per item	Wool used per item	Labor-hours per item	Number made
Coat	$300	2	4	4	x
Dress	$140	4	0	2	y
		$5 per sq yd; 120 sq yd available	$20 per sq yd; 100 sq yd available	$25 per labor-hour; 110 labor-hours available	

Since net income is to be maximized we turn our attention to expressing net income in terms of x and y.

Net income = Amt from sales—Production cost

$$I(x,y) = \underbrace{300x + 140y}_{\text{sales}} - \underbrace{\left[5(2)x + 20(4)x\right]}_{\substack{\text{cost of coat} \\ \text{material}}} - \underbrace{5(4)y}_{\substack{\text{cost of} \\ \text{dress} \\ \text{material}}}$$

$$-\underbrace{\left[25(4)x + 25(2)y\right]}_{\text{labor cost}}$$

By multiplying and collecting terms we obtain:

$$I(x, y) = 110x + 70y$$

The constraint $4x \leq 100$ or, equivalently, $x \leq 25$, expresses the condition that the amount of wool used cannot exceed 100 square yards; $2x + 4y \leq 120$ expresses the condition that the amount of cotton used cannot exceed 120 square yards; $4x + 2y \leq 110$ expresses the condition that the number of labor-hours employed cannot exceed 110.

We emerge with the following linear program:

$$\text{Maximize } I(x, y) = 110x + 70y$$

subject to
$$x \geq 0, y \geq 0$$
$$x \leq 25$$
$$2x + 4y \leq 120$$
$$4x + 2y \leq 110,$$

which with the underlying **assumptions/postulates** stated in (b) is the L.P. model that emerges.

(d)   The graph of the feasible points is shown in Figure 2.3. The corner

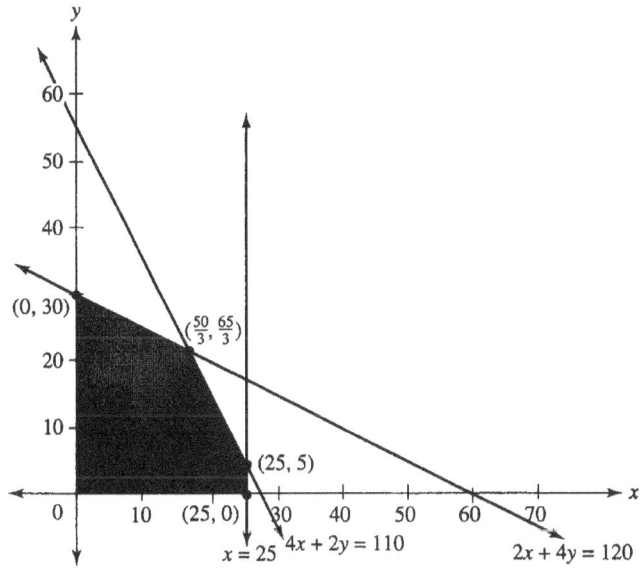

Figure 2.3

point $\left(\dfrac{50}{3}, \dfrac{65}{3}\right)$ yields the maximum value 3350.

(e)   Suppose review of the **assumptions/postulates** in (b) finds them realistic, but implementation of the afore solution is not possible.

(g) To be implemented sometimes the L.P. model requires a
     solution in integers. Application of the corner point method
     does not always yield a solution in integers, as in the current
     problem. Methods for obtaining the best solution in integers
     are needed for such problems. In this case (17, 21) is the
     solution in integers and 3340 is the maximum value.

6. Certitude?

Disagree. Mathematical methods have the advantage of certitude in the sense of validity, not truth. The statement confuses truth with validity. Replace truth where it appears with validity and we have a correct assertion.

2.9 LINEARITY *ÜBER ALLES*?

Advertising Media Selection

The general advertising media-selection problem is to choose from various media capable of carrying an advertisement a selection that is, in some sense, best. Specific choices within a given medium as well as given media are included in the alternatives. The constraints in media selection include the size of the advertising budget, the minimum and maximum usages of specific media categories, and the desired minimum exposure rate to envisioned buyers. A number of approaches have been developed for the media-selection problem[3], and in the early 1960's hopes ran high in the world of advertising for the use of linear programming. An early linear-program model for media selection was the one developed by James Engel and Martin Warshaw[4] for the McGraw-Edison Company. The Pennsylvania Transformer division of McGraw-Edison manufactures transformers for use by industrial plants, schools, hospitals, commercial construction projects, and so on. Ten trade publications were considered for advertising purposes, and $25,000 was allocated for industrial advertising for a period of 1 year. Since the purchase decision is usually made by the plant engineer, the objective posed was to maximize the number of plant engineers reached. The following linear program model was developed:

---

[3] See Dennis Gensch, "Different Approaches to Advertising Media Selection," *Operational Research Quarterly*, vol. 21, no. 2 (June 1970), pp. 193-219; Philip Kotler, *Marketing Management* (Englewood Cliffs, N.J.: Prentice-Hall, Inc., 1967), Chapter 18.

[4] "Allocating Advertising Dollars by Linear Programming," *Journal of Advertising Research*, vol. 4, no. 3 (September 1964), pp. 42-48.

Maximize $f = 0X_1 + 25X_2 + 32.87X_3 + 49.44X_4 + 56.65X_5 + 17.54X_6 + 58.20X_7 + 0X_8 + 23.53X_9 + 40.00X_{10}$

subject to nonnegativity of the variables ($X_1 \geq 0$, $X_2 \geq 0$, etc.)

$$X_1 \leq 5,400$$
$$X_2 \leq 9,504$$
$$X_3 \leq 8,760$$
$$X_4 \leq 10,680$$
$$X_5 \leq 11,016$$
$$X_6 \leq 5,472$$
$$X_7 \leq 9,072$$
$$X_8 \leq 3,300$$
$$X_9 \leq 8,160$$
$$X_{10} \leq 6,900$$
$$X_1 + X_2 + X_3 + X_4 + X_5 + X_6 + X_7 + X_8 + X_9 + X_{10} \leq 25,000$$

where X_1 is the amount to be invested in media 1 (*Consulting Engineer Magazine*), X_2 is the amount to be invested in media 2 (*Electrical Construction Magazine*), and so on. The coefficients 0 of X_1, 15.15 of X_2, 32.87 of X_3, and so on, in the linear function f represent the number of plant engineers reached by each magazine per advertising dollar invested, so that f represents the total number of plant engineers reached. The last constraint expresses the condition that no more than $25,000 is to be spent on advertising and the other constraints are to prevent more dollars from being invested in any one monthly magazine than is necessary to buy 12 insertions.

Although linear program models were satisfactory for crude versions of the media-selection problem, it soon became clear that the features exhibited by more sophisticated versions of the problem could not be modeled realistically in linear programming terms.

Frank Bass and Ronald Lonsdale[3] found linear-program models to be crude devices to apply to the media-selection problem. The **linearity assumption** itself, is the source of much of the difficulty. Justifying an **assumption of linear response** to advertising exposures on theoretical grounds would be difficult. **Assumptions** about the nature of response to advertising cause most difficulties in models of the type examined in this article.

Philip Kotler[4] noted the following limitations:

> Linear programming **assumes** that repeat exposures have a constant marginal effect.
>
> It **assumes** constant media costs (no discounts).
>
> It cannot handle the problem of audience duplication.
>
> It says nothing about when ads should be scheduled.

[6]

Although later linear programming approaches to the media-selection problem sought to overcome the criticisms that had been voiced, the message was clear: although linearity, as a mathematical tool, is too good not to be true, a linear-program model is not always a suitable fit for a media-selection problem; that is, the **assumptions** that must be made to force a fit are not always sufficiently realistic.

[5]   "An Exploration of Linear Programming in Media Selection," *Journal of Marketing Research*, vol. III, no. 2 (May 1966), pp. 179-188.

[6]   *Marketing Management* (Englewood Cliffs, N.J.: Prentice-Hall, Inc.) p. 478.

3

Are Leontief Input - Output Models Realistic?

3.1 PREFACE

Input-Output Models for economic systems were pioneered by the economist Wassily Leontief, a recipient of the 1973 Nobel Prize in Economics. They are often discussed as an application in finite math courses that take up matrix algebra, but the sample of finite math books I examined said nothing about the **assumptions** that underlie them.

3.2 REVIEW OF THE FUNDAMENTALS

In input-output analysis an economic system is viewed as a collection of interacting industries in which each industry produces an output that serves as raw materials, or input, for the industries of the system and requires input from the industries of the system. Let a_{ij} denote the amount of input (dollar's worth) of commodity i needed to produce \$1 worth of commodity j; the first subscript refers to input, the second to output. Thus, for example, the equation $a_{21} = 0.20$ asserts that 20¢ worth of commodity 2 is needed to produce \$1 worth of commodity 1. For an n-industry economy, the matrix

$$
A = \begin{bmatrix}
a_{11} & a_{12} & \cdots & a_{1n} \\
a_{21} & a_{22} & \cdots & a_{2n} \\
\cdot & & & \cdot \\
\cdot & & & \cdot \\
\cdot & & & \cdot \\
a_{n1} & a_{n2} & \cdots & a_{nn})
\end{bmatrix},
$$

called the **input-coefficient matrix of the system,** specifies the amount of each commodity that is needed to produce $1 worth of each commodity. The entries in the first column, for example, specify the inputs required from each of the n industries to produce $1 worth of the commodity produced by industry 1.

The entries in the first row specify the amount of the commodity provided by industry 1 needed to produce $1 worth of the commodities produced by the n industries of the system.

We also **assume** that there is an **open sector in the economy** (consisting of households, for example) that absorbs a noninput demand for the product of each industry and supplies the primary input, labor. Let d_1, d_2,..., d_n denote the demand of the open sector for the commodities produced by industries 1, 2,..., n and let x_1, x_2,..., x_n denote the total output (dollar's worth) of industries 1, 2,..., n.

The product $a_{ij}x_j$ is (the amount of commodity i needed to produce $1 worth of commodity j) x (total dollar's worth of commodity j produced), and thus expresses the input requirement of industry j for commodity i. For example, if a_{ij} = 0.20 (20¢ worth of commodity i is needed to produce $1 worth of commodity j) and x_j = 5000 ($5000 worth of commodity j is produced), then 0.20(5000) = $1000 worth of commodity i is needed to produce commodity j.

Thus $a_{11}x_1$ is the input requirement of industry 1 for commodity 1, $a_{12}x_2$ is the input requirement of industry 2 for commodity 1, $a_{13}x_3$ is the input requirement of industry 3 for commodity 1, and so on. The sum

$$a_{11}x_1 + a_{12}x_2 + \dots + a_{1n}x_n + d_1$$

is the sum of the input requirements of the n industries and the open sector for commodity 1. For x_1, the total output of industry 1, to satisfy this demand, we must have:

$$x_1 = a_{11}x_1 + a_{12}x_2 + \dots + a_{1n}x_n + d_1$$

Similarly, for x_2, the total output of industry 2, to satisfy the demand for commodity 2, we must have:

$$x_2 = a_{21}x_1 + a_{22}x_2 + \ldots + a_{2n}x_n + d_2$$

More generally, for x_n, the total output of industry n, to satisfy the demand for commodity n, we must have:

$$x_n = a_{n1}x_1 + a_{n2}x_2 + \ldots + a_{nn}x_n + d_n$$

Thus the conditions that must be satisfied by output levels $x_1, x_2, \ldots, x_n$, of the n industries in the economy to satisfy the demands of the open sector and the industries themselves are expressed by the following system of n equations:

$$x_1 = a_{11}x_1 + a_{12}x_2 + \ldots + a_{1n}x_n + d_1$$
$$x_2 = a_{21}x_1 + a_{22}x_2 + \ldots + a_{2n}x_n + d_2$$
$$\vdots \qquad \vdots$$
$$x_n = a_{n1}x_1 + a_{n2}x_2 + \ldots + a_{nn}x_n + d_n$$

Rewriting this system so that terms involving $x_1, x_2, \ldots, x_n$ appear on one side and the constants $d_1, d_2, \ldots, d_n$ appear on the other side yields:

$$(1 - a_{11})x_1 - a_{12}x_2 - \ldots - a_{1n}x_n = d_1$$
$$- a_{21}x_1 + (1 - a_{22})x_2 - \ldots - a_{2n}x_n = d_2$$
$$\vdots \qquad \vdots$$
$$- a_{n1}x_1 - a_{n2}x_2 - \ldots + (1 - a_{nn}x_n) = d_n \qquad (3.1)$$

It is advantageous to express this system in terms of a matrix equation involving a matrix product. To do so we introduce matrices, I_n, X, and D, as follows:

$$I_n = \begin{bmatrix} 1 & 0 & 0 & \cdots & 0 \\ 0 & 1 & 0 & \cdots & 0 \\ 0 & 0 & 1 & \cdots & 0 \\ \cdot & \cdot & \cdot & & \cdot \\ \cdot & \cdot & \cdot & & \cdot \\ \cdot & \cdot & \cdot & & \cdot \\ 0 & 0 & 0 & \cdots & 1 \end{bmatrix}, \quad X = \begin{bmatrix} x_1 \\ x_2 \\ \cdot \\ \cdot \\ \cdot \\ x_n \end{bmatrix}, \quad D = \begin{bmatrix} d_1 \\ d_2 \\ \cdot \\ \cdot \\ \cdot \\ d_n \end{bmatrix}$$

Matrix I_n is an n by n matrix with 1's in the main diagonal and 0's elsewhere; X is called the **output matrix** of the system and D is called the **final-demand matrix** of the system.

Subtracting A, the input-coefficient matrix, from I_n yields:

$$I_n - A = \begin{bmatrix} (1-a_{11}) & -a_{12} & \cdots & -a_{1n} \\ -a_{21} & (1-a_{22}) & \cdots & -a_{2n} \\ \cdot & & \cdot & \\ \cdot & & \cdot & \\ \cdot & & \cdot & \\ -a_{n1} & -a_{n2} & \cdots & (1-a \end{bmatrix}$$

Taking the product of $(I_n - A)$ and X, $(I_n - A)X$, yields the left side of system (3.1); the right side of system (3.1) is expressed by matrix D. Thus, in matrix terms, system (3.1) is expressed by the following matrix equation:

$$(I_n - A)X = D$$

In summary, then, the problem of satisfying the needs of the n industries of the economy, expressed by the input-coefficient matrix A, and the needs of the open sector of the system, expressed by the final-demand matrix D, reduces to the matrix problem of determining output matrix X such that the product $(I_n - A)X$ equals D.

As long as the input-coefficient matrix A does not change, $(I_n - A)^{-1}$ does not change, and with one matrix inversion a variety of possible final-demand situations can be studied.

The successful application of the input-output model to an economy requires that a **realistic input-coefficient matrix A and final demand matrix D** be developed for the economy.

Assuming that this could be done, the second major problem is to suitably refine these matrices so that they are **realistic over time**.

3.3 BEWARE THE ASSUMPTIONS

The afore **assumption** is not one that should be taken lightly. Hollis Chenery and Paul Clark [2] and William Baumol [1] note:

The properties of Leontief models can be derived from **three basic assumptions**, which it will be useful to state at the outset:

(1)　*Each commodity (or group of commodities) is supplied by a single industry or sector of production.* Corollaries of this **assumption** are (*a*) that only one method is used for producing each group of commodities; and (*b*) that each sector has only a single primary output.

(2)　*The inputs purchased by each sector are a function only of the level of output of that sector.* (The **stronger assumption** is usually made that the input function is linear, but this is a matter of convenience.)

(3)　*The total effect of carrying on several types of production is the sum of the separate effects.* This is known as the **additivity assumption**, which rules out external economies and diseconomies.

The validity [meaning realism, here] of each of these **assumptions** depends both on the nature of production in single plants and on the way in which these units are aggregated into sectors. Some **assumptions** may be more valid [that is, realistic] for aggregates than for individual units, as for example the exclusion of joint products and external economies. Others may hold for single productive processes but not for sectors. We must therefore consider together the nature of the underlying production relationships and the effects of aggregation in evaluating the structure of the model. [2; 33-4]

Perhaps more serious is a **second assumption** which states that in any productive process all inputs are employed in rigidly fixed proportions and the use of these inputs expands in proportion with the level of output. This is a special case of an **assumption of constant returns to scale** (see Chapter 11, Section 5). But the **fixed-proportions assumption** is far more restrictive. Constant returns to scale is perfectly consistent with the substitution of one factor for another. A linear homogeneous production function (constant return to scale) permits both labor-intensive and capital-intensive processes. The firm whose production function exhibits constant returns can if it wishes have one hundred workers for every $1,000 invested in machinery, or it may use machines which require only ten workers per $1,000 machine investment. A linear homogeneous production function requires only that if the firm decides to triple the scale of either of these types of operation, the result will be a tripling of output. Not so the Leontief fixed-proportions premise, which requires that a manufacturing process which is labor intensive offer no option of a capital-intensive alternative. If fifty-three men per $1,000 of investment are required at any level of operation, it is **assumed** that the same ratio will be required no matter how much the size of the firm expands or contracts. Whether this **assumption** is relatively innocuous or does considerable violence to the input-output results is still under dispute. But the premise is certainly never absolutely true, even in those cases where chemistry and engineering dictate fixed proportions between some ingredient and output. [1; 538-9.]

References

[1]   W.J. Baumol. *Economic Theory and Operations Analysis,* 4th ed. Englewood Cliffs, N.J.: Prentice-Hall, Inc., 1977, Chapter 22.

[2]   H.B. Chenery and P.G. Clark. *Interindustry Economics.* New York: John Wiley & Sons, Inc., 1959.

[3]   Conference on Research in Income and Wealth, National Bureau of Economic Research. *Input-Output Analysis: An Appraisal.* Princeton, N.J.: Princeton University Press, 1955.

[4]   R. Dorfman, P.A. Samuelson, and R.M. Solow. *Linear Programming and Economic Analysis.* New York: McGraw-Hill Book Company, 1958, Chapters 9-12.

[5]   W.W. Leontief. *The Structure of American Economy, 1919-1939,* 2d ed. New York: Oxford University Press, 1951.

[6]   W.W. Leontief, ed. *Studies in the Structure of the American Economy.* New York: Oxford University Press, 1953.

3.3   FOOD FOR THOUGHT

1.   A Two-Industry Economy

For a two-industry economy, let us **assume** that $0.20 and $0.30 worth of the first industry's commodity is needed by the first and second industries to produce $1 worth of their respective commodities and that $0.30 and $0.10 worth of the second industry's commodity is needed by the first and second industries to produce $1 worth of their respective commodities.

(a)   Set up the input-coefficient matrix of the economy.

(b)   If the open sector of the economy requires $2520 worth of commodity 1 and $3150 worth of commodity 2, what output levels will satisfy the input needs of the industries and the requirements of the open sector?

How much of these outputs will be consumed by industries 1 and 2?

(c) By itself the afore is an exercise in generating matrices and numbers. Are there underlying conditions that make it worth something more? How so?

3. 4 ANSWERS/DISCUSSION OF FOOD FOR THOUGHT

1.   **A Two-Industry Economy**

(a) $A = \begin{bmatrix} 0.2 & 0.3 \\ 0.3 & 0.1 \end{bmatrix}$, $D = \begin{bmatrix} d_1 \\ d_2 \end{bmatrix}$, and

$X = \begin{bmatrix} x_1 \\ x_2 \end{bmatrix}$ are the input-coefficient, final demand, and output matrices, respectively. We have:

$$I_2 - A = \begin{bmatrix} 0.8 & -0.3 \\ -0.3 & 0.9 \end{bmatrix}, \ (I_2 - A)^{-1}$$

$$= \begin{bmatrix} \dfrac{90}{63} & \dfrac{30}{60} \\ \dfrac{30}{63} & \dfrac{80}{63} \end{bmatrix}, \ x = (I_2 - A)^{-1} D$$

$$= \begin{bmatrix} \dfrac{90}{63} d_1 + \dfrac{30}{60} d_2 \\ \dfrac{30}{63} d_1 + \dfrac{80}{63} d_2 \end{bmatrix}$$

(b) $ 5100 worth of commodity 1 and $ 5200 of commodity 2; $ 1020 worth of commodity 1 is needed to produce commodity 1, and $1560 worth of commodity 1 is needed to produce commodity 2; $1550 worth of commodity 2 is needed to produce commodity 1, and $520 worth of commodity 2 is needed to produce commodity 2.

(c)   Worth something more translates to realistic description of the functioning of the economy. This requires that the input-coefficient matrix A and final demand matrix D be realistic. **Beware the Assumptions**.

4

Equally-Likely Outcome Probability Models *Über Alles,* No Ifs-ands-or Buts?

4.1 PREFACE

Quote from a textbook in the sample I gave thought to:

Basic Probability Principle

Let S be a sample space of equally likely outcomes and let E be a subset of S. Then the probability that event E occurs is

$$P(E) = \frac{n(E)}{n(S)}$$

where n(E) is the number of outcomes in E and n(S) is the number of outcomes in S.

There is a But in the form of a Caution. The basic probability principle only applies when the outcomes are equally-likely, the meaning of which is not discussed. The discussion of probability that follows is based on the afore basic probability principle, no exceptions, so that it is not necessary to give thought to the Caution.

Although details vary, in the finite math books I examined over the years the equally-likely outcome approach to probability in terms of focus, illustrations, and homework assignments is common to all.

4.2   Student Replies to Questions

What is the probability of dealing a picture card from a standard deck of 52 cards?, I ask my finite math classes. It's $\frac{12}{52}$, I'm told, for which there is general agreement with a few students not sure or not willing to commit themselves.

I take a die from my pocket, toss it on my desk, and ask for the probability of an even number showing top side. ½, I'm told.

A poker hand of 5 cards is to be dealt from a standard deck of 52 cards. What is the probability that it contains 3 kings, I ask. For students whose math background includes combinatorial methods the answer is:

$$\frac{C(4,3) \cdot C(48,2)}{C(52,5)}$$

4.3   Unsatisfactory Introduction to Probability?

An introduction to probability through

$$P(E) = \frac{n(E)}{n(S)}$$

as noted in the preface, is unsatisfactory in that it is a version of It's-Just-a-Word-Problem approach with the perspective that the probability conclusion reached is a real-world fact, period, rather than a valid conclusion of an underlying math (probability) model.

5

An Approach to Probability Modeling which Puts the Equally-Likely Outcome Model in Proper Perspective

5.1   PREFACE

Objective: state as background properties of the relative frequency of occurrence of an event, call it A, defined by

$$R(A) = \frac{\text{number of times the process occurs}}{\text{number of times A occurs}}$$

and use this background to develop the concept of (finite) probability model.

5.2   PROPERTIES OF RELATIVE FREQUENCY

1. $R(A) \geq 0$. Both numerator and denominator of $R(A)$ are nonnegative.

2. $R(A) \leq 1$. The numerator of $R(A)$ cannot exceed its denominator.

3. $R(A) = 0$, if A is an event whose occurrence is not possible (which happens when A is defined by incompatible conditions). Such an event A is identified with $\emptyset$.

4. $R(S) = 1$, where $S = \{s_1, s_2,..., s_n\}$, is a sample space for the random process. Whenever the process is repeated one of the sample points in S occurs, and thus S occurs. Therefore, the numerator and denominator of $R(S)$ are equal.

5. $R(A)$ = sum of the relative frequencies of the sample points that describe A.

6. $R(S) = R(s_1) + R(s_2) + ... + R(s_n)$, where $S = \{s_1, s_2,..., s_n\}$ is a sample space for the process. This property is obtained by applying property 5 to S.

7. $R(s_1) + R(s_2) + ... + R(s_n) = 1$; the sum of the relative frequencies of all sample points is 1. This follows from properties 4 and 6.

5.3 THE CONCEPT OF (FINITE) PROBABILITY MODEL

To define the concept of (finite) probability model follow in the footsteps of properties of relative frequency.

A **(finite) probability model** for a random process consists of two components:

(i)　A **sample space** $S = \{s_1, s_2,..., s_n\}$.

(ii)　A function P, called a **probability function**, which assigns to each subset A of S a value, denoted by $P(A)$ and called the probability of A, subject to the conditions listed below. We state these conditions on the sample points and then extend them to subsets of S.

1.　$P(s_1) \geq 0, P(s_2) \geq 0,..., P(s_n) \geq 0$

2.　$P(s_1) \leq 1, P(s_2) \leq 1,..., P(s_n) \leq 1$

3.　$P(s_1) + P(s_2) + ... + P(s_n) = 1$

4.　If A is a subset of S, $P(A)$ = sum of the probabilities of the sample points describing A; if $A = \emptyset$, $P(A) = 0$. (For example, if $A = \{s_1, s_5\}$, $P(A) = P(s_1) + P(s_5)$.)

I employ the following analogy which, in my experience, my students find helpful.

Building Code

The structural requirements of a probability model for a random process are analogous to a community's building code requirements for building a house. A building code does not tell us how to build a house. It tells us that however we build our house for it to be legitimate in terms of the building code, it must satisfy such and such conditions which are spelled out in the code.

The concept of probability model does not tell us how to define a sample space and probability function for a random process; it tells us that in building a specific probability model for a random process, which can be done in many ways, we must satisfy the conditions stated in the definition of probability model (the building code in this case) in order for the model to be mathematically legitimate.

Mathematical Legitimacy: Janet James's Probability Model

Janet James was presented with the following structure which was claimed to be a probability model for the process of tossing a die.

$$S = \{O, E\}, P(O) = \frac{1}{3}, P(E) = \frac{2}{3},$$

where O is the event that an odd number shows and E is the event that an even number shows on a throw of the die. Is this structure a probability model?

S is a sample space for the die tossing process since it satisfies the requirement that exactly one of its events occur when the die is tossed. The assignment by P of the numerical values to O and E satisfies the requirements of a probability function since 1/3 and 2/3 are non-negative, less than 1, and sum to 1. This establishes the **mathematical legitimacy** of $\{S, P\}$ as a probability model for the die tossing process; whether or not this model **realistically** describes the behavior of any die in Janet's possession is another question entirely.

I find that it supports clarity to emphasize the following:

Probability Model Properties vs. Relative Frequency Properties

At first sight, I point out, the concept of probability model seems no different from properties of relative frequency.

This is not the case, I emphasize. The conditions required of a probability function closely mirror properties of relative frequency, but relative frequency is defined in a very specific way while probabilities can be assigned to events in a wide variety of ways as long as the conditions cited are satisfied. Probability assignments reflect properties of relative frequency, but go beyond them in much the same way that a son may reflect properties of his father, but goes beyond them.

5.4   THE SPECIAL CASE OF EQUALLY-LIKELY OUTCOME PROBABILITY MODELS

Basic Properties

> **Theorem 1.** Let $S = \{s_1, s_2,..., s_n\}$ denote a sample space with n sample points. Let us suppose that for one reason or another we are led to the probability function P which assigns the same value x to all of the sample points in S; then this value is $x = \dfrac{1}{n}$.

Proof: From the definition of probability model, we have:

$$P(s_1) + P(s_2) + ... + P(s_n) = 1$$

Since each of $P(s_1)$, $P(s_2)$,..., $P(s_n)$ equals x, we obtain:

$$\underbrace{x + x + \ldots + x}_{n \text{ terms}} = 1$$

$$nx = 1$$

$$x = \frac{1}{n}$$

A probability model in which all sample points are assigned the same probability value is called an **equally-likely outcome model**.

Theorem 2. If $S = \{s_1, s_2, \ldots, s_n\}$ $P(s_1) = \ldots = P(s_n) = 1/n$, and A is an event that is described by k sample points, then

$$P(A) = \frac{k}{n}.$$

Proof: For the sake of simplifying our discussion, let us suppose that A is described by the first k sample points $s_1, s_2, \ldots, s_k$. Then we have

$$P(A) = P(s_1) + P(s_2) + \ldots + P(s_k)$$

$$= \underbrace{\frac{1}{n} + \ldots + \frac{1}{n}}_{k \text{ terms}} = \frac{k}{n}$$

In the special case of equally-likely outcomes probability questions reduce to counting questions. To determine the probability of A in this framework, count the number of sample points that describe A, count the number of sample points, and take their ratio.

If S contains 10,000 sample points and A is described by 1000 of them, then, in this model,

$$P(A) = \frac{1000}{10,000} = \frac{1}{10}.$$

The essential qualification in the afore which I always emphasize for my students is **"in this model"**.

6

Which Is the "Right" Probability Model?

6.1 Preface

Please recall from Chapter 2 that "Right" means in the sense of being realistic, which is the sense in which it is used in this setting.

I invite you to consider the following seven situations.

6.2 Case 1: Dealing a Card from a Standard Deck of 52 Cards

I find this case a useful one to return to because of the usual student answer $\frac{12}{52}$ (no if's-ands-or-buts) that I receive when I ask, what is the probability of dealing a picture card from a standard deck of 52 cards? (see Sec. 4.2.) It provides an opportunity to discuss why there's much more to it than that.

Consider the process of dealing a card from a standard deck of 52 cards and let us address the problem of setting up a probability model for this process and determining the probability that a picture card is dealt.

For convenience in referring to the cards, let us set up a translation system so that we can refer to the cards as 1, 2, . . . , 52; in this translation system. 1 might denote the ace of spades, 2 the king of spaces, etc. We take as our sample space,

$$S = \{C_1, C_2,..., C_{52}\},$$

where C_1 is the event that card 1 is dealt, C_2 is the event that card 2 is dealt, etc.

Our next task is to define a probability function P on the subsets of S. This can be done in many ways, and the function P that emerges depends on the **assumption** we make about how the card will be dealt from the deck.

Suppose we **assume** what is usually assumed in such situations, but not always made explicit, that the card is dealt from a well-shuffled deck in an unbiased way—at random, as we say. The probability function P which best reflects this **assumption** assigns the same value, 1/52, to each sample point in S.

We thus emerge with the following probability model:

$$\textbf{Model 1}: S = \{C_1, C_2,..., C_{52}\}$$

$$P(C_1) = ... = P(C_{52}) = \frac{1}{52}$$

From Model 1 it follows that the probability that a picture card is dealt is $\frac{12}{52} = 0.23$.

As is well known, some card dealers are less than honest. Suppose we **assume**, based on past experience, that the dealer intends to "arrange things" so that the card we are dealt is neither a picture card nor an ace, but that any of the other cards may be dealt without bias. For notational convenience suppose that the picture cards and aces are in the cards we numbered 1,2, . . . 16, and that the cards 17, 18, . . . 52 correspond to the remaining cards. This leads to Model 2.

$$\textbf{Model 2}: S = \{C_1,... C_2,..., C_{52}\},$$

$$P(C_1) = ... = P(C_{16}) = 0, P(C_{17}) = ... = ... = P(C_{52}) = \frac{1}{36}$$

From Model 2 it follows that the probability that a picture card is dealt is 0.

Which model is "correct"? Correct is a word that we must always consider with caution because of its double edged nature.

In the sense of satisfying the conditions that define a probability model, both are correct. The conclusions obtained from these models are correct in the sense of being valid with respect to each of them.

Which model, if either, is correct in the sense of being realistic? Only real-world experience with cards can help us address this question.

6.3   CASE 2: THE ADVENTURES OF HASTY HARRY

I put in the form of a story situations that Harry finds himself to probe the distinction between validity and realism. The story provides an opportunity for food-for-thought for students which, in a simple setting, probe the issues that are part of more complex settings.

What should I get my brother for his birthday, pondered bottom-line Bob. It's his thirtieth, the big 30, and this calls for something very special. His brother Harry, who was fond of games of chance, had an extensive collection of "unusual" dice. Bob decided to add to it by obtaining for him what would undoubtedly be the crown jewel of his collection, a gold die embedded with diamond chips to show off the spots on its faces. Bob had the die custom made and on the appointed day a very pleased Harry received a very special die.

Harry could hardly wait to show off his new treasure to his friends and, in addition, win some vacation money in a bit of friendly gaming activity that was certain to follow. In preparation for this he went to Martin's Models to obtain a probability model to describe the behavior of the new crown jewel of his collection.

"Mr. Martin, I want you to build me a probability model for the tossing of this die. I'm particularly interested in the probability that an even number shows."

"What can you tell me about your die, Harry? What do you know about its behavior from your experience with it?"

"I just got it as a present and I have no experience with it. I want to be prepared with a probability model before obtaining that experience. A die is a die, nothing special, apart from its being made of gold with diamond chips. What else is there to know?"

"All right Harry, I'll proceed on the **assumption** that it's an ordinary die of uniform composition, a fair die, as we say. This being the case the equally likely outcome model R4 (red) that I have in stock is the most realistic description of the behavior of your die.

R4 (red): $S = \{1, 2, 3, 4, 5, 6,\}$

$$P(1) = P(2) = ... = P(6) = \frac{1}{6}$$

It follows from R4 that the probability that an even number shows is:

$$P(2) + P(4) + P(6) = \frac{3}{6}$$

$$= 0.50$$

"What does this mean in terms of some friendly gaming activity, Mr. Martin?"

"If you toss this die a large number of times, an even number should show roughly half the time. We cannot say when an even number will show, that's a matter of chance, but it should show roughly 50% of the time."

Reality Strikes

Harry proudly showed his die to his friends and all, with the exception of Harry, had a good time playing games of chance in which "friendly" bets were placed on which face would show when the die was tossed. Harry expected an even number to show roughly 50% of the time as predicted by Martin's model and bet accordingly. It came to pass, however, that after 500 tosses of his die an even number had showed 66% of the time, which is sharply at variance with what Harry had expected. He had hoped to make a modest profit from this "friendly" gaming activity and now he found himself an unfriendly three hundred bucks in the red. Confused and feeling that he had been cheated, he stormed back to Martin's models for some answers.

"I don't understand what went wrong Martin. If your mathematics is so precise, how could it happen that an even number showed 66% of the time instead of around the 50% you told me to expect? I'm three hundred bucks down because of this. You sold me a defective model and I want my money back."

"Mathematics, the probability model I gave you in this case, Harry, delivered what it was capable of, namely, a **valid conclusion with respect to the assumptions made.** Please remember that it was you who provided me with the starting point of the analysis. I quote: 'a die is a die, nothing special.' As it turned out there was something very special about this die which **made my assumption, based on your information, unrealistic.** As a result we obtained a valid conclusion from the model which, when interpreted in relative frequency terms, tells us about how often an even number can be expected to show. This proved to be at variance with the nature of your die.

When there is a sharp conflict between a math model and the reality it is intended to describe, **reality always wins.**

Didn't your brother say anything to you about the nature of the die, or were you too dazzled by the gold and diamond chips to pay attention?"

"I'm not sure now. I'll have to ask him. Maybe I was too hasty."

A New Model for Hasty Harry's Die

"Mr. Martin, I spoke to my brother and I listened this time. He said he told me that he had the die weighted internally so that the even numbered faces were twice as likely to show as the odd numbered ones. It's not at all an ordinary die in terms of its internal make up."

"I spoke to Bob after you left and he told me about the die's structure. He spent a lot of money to have the die made in this way and, ironically, it ended up costing you money. I developed another probability model for your die based on the information Bob gave me. This model should be a much better fit to your die.

I call it B7 (blue); it is defined by,

$$B7 \text{ (blue) } S = \{1, 2, 3, 4, 5, 6\}$$

$$P(1) = P(3) = P(5) = \frac{1}{9}$$

$$P(2) = P(4) = P(6) = \frac{2}{9}$$

It follows as a valid conclusion from B7 that the probability an even number shows is:

$$P(2) + P(4) + P(6) = \frac{2}{9} + \frac{2}{9} + \frac{2}{9}$$

$$= \tfrac{2}{3} = 0.666$$

The relative frequency interpretation of this valid conclusion is that if your die were tossed a large number of times, an even number would show about 67% of the time. This, I understand, is in agreement with the "friendly" game experience that you had."

"How much do I owe you?"

"The same as for the previous model. There's no additional charge for your hasty action. You've paid that price."

Hasty Harry's Other Die

"I have another die in my collection Mr. Martin, one that is weighted in such a way that the even numbered faces are three times as likely to show as the odd numbered ones. Do you have a suitable model for this die?" "As a matter of fact I do. It's Y3 (yellow) in my catalog listing and it's defined by:

$$Y3 \text{ (yellow) } S = \{1, 2, 3, 4, 5, 6\}$$

$$P(1) = P(3) = P(5) = \frac{1}{12}$$

$$P(2) = P(4) = P(6) = \frac{3}{12}$$

6.4 FOOD FOR THOUGHT

1.   **Harry's Other Die**

Consider the yellow model Y3 for the process of tossing Harry's other die.

$S = \{1, 2, 3, 4, 5, 6\}$,

$$P(1) = P(3) = P(5) = \frac{1}{12}; \qquad P(2) = P(4) = P(6) = \frac{3}{12}.$$

Let E denote the event that an even number shows.

(a)   Find $P(E)$.

(b)   State the relative frequency interpretation of the result obtained in (a).

(c)   In tossing the other die in Harry's collection 1000 times, an even number was observed to show 665 times. Does this show that the conclusion obtained in (a) is not valid?

(d)   Is the conclusion reached in (a), interpreted in relative frequency terms, realistic?

(e) Is the yellow model Y3 realistic for the afore die?

6.5 ANSWERS/DISCUSSION OF FOOD FOR THOUGHT

1. Harry's Other Die

(a) $P(E) = P(2) + P(4) + P(6) = 3/4 = 0.75$. (b) If a die whose behavior is described by model Y3 is tossed a large number of times, an even number will show approximately 75% of the time. (c) The tossing of the die a large number of times and observing how often an even number shows is **irrelevant** to the **validity issue.** The **validity** of $P(E) = 0.75$ was established by adding up the probabilities with which 2, 4 and 6 show in model Y3 and obtaining 0.75.

(d) From tossing the die a large number of times, we have that the relative frequency with which an even number showed is 0.665, which is markedly at variance with the predicted relative frequency of approximately 0.75 in (b). This establishes that $P(E) = 0.75$, interpreted in relative frequency terms, is **false.**

(e) Model Y3 is **not realistic** for the die in question since **a valid conclusion of the model has been shown to be false.**

6.6 CASE 3: A PRODUCTION PROCESS PROBLEM

The production process problem serves as a vehicle for introducing a number of important issues which are further explored in food-for-thought.

The Bokson Company makes television tubes in two plants, B1 and B2. The weekly output is 3000 tubes, with 1800 tubes being produced in plant B1, of which 1 percent are defective, and 1200 tubes being produced in plant B2, of which 2 percent are defective.

A tube is selected **at random** from the week's output. The problem is to set up a probability model for this selection process and find the probability that a defective tube is chosen.

We take as our sample space the set of 3000 events $S = \{t_1, t_2,..., t_{3000}\}$, where t_1 is the event that tube 1 is chosen, t_2 is the event that tube 2 is chosen, and so on. To say that a tube is chosen at random means that a tube is chosen in an unbiased way so that certain tubes are not favored over others. In other words, all tubes have the same likelihood of being chosen. This **assumption** is best reflected by the probability function P defined by:

$$P(t_1) = P(t_2) = ... = P(t_{3000}) = \frac{1}{3000}$$

To obtain the probability that a defective tube is chosen in terms of this probability model, we must find the number of ways of selecting a defective tube and divide by 3000, the total number of sample points. Since 1 percent of the output of plant B1 is defective, B1 contributes $0.01(1800)$ or 18 defectives to the total number of defectives. Plant B2 contributes $0.02(1200)$ or 24 defectives, and the total number of defectives is 42. Thus:

$$P(\text{defective tube is chosen}) = \frac{42}{3000} = 0.014$$

The relative-frequency interpretation of this result is that, if the envisioned selection process is performed a large number of times, a defective tube will be chosen approximately 1.4 percent of the time.

> The basic **assumption** that led to this model is that the drawing of a tube is made **at random**; that is, in an unbiased way, no favoritism, deliberate or inadvertent.
>
> This is much more easily said than done. If the model constructed is to be realistic for the drawing, then we must do everything possible to ensure that the randomness assumed in theory is lived up to in practice.

Suppose that in carrying out the selection process as envisioned a large number of times a defective tube was chosen 5 percent of the time.

> What does this tell us? It tells us that the random sampling envisioned in theory was not achieved in practice.

6.7 FOOD FOR THOUGHT

1. **The Twolow Company**

    The Twolow Company makes light bulbs. Two plants, $P1$ and $P2$, carry out the production process. The daily output is 8000 bulbs, with $P1$ producing 5000 bulbs of which 1 percent are defective, and $P2$ producing 3000 bulbs of which 0.5% are defective.

    A bulb is selected from the day's output.

    (a) When asked to determine the probability that a defective bulb is chosen, Mark Twolow set up the following probability model for the selection process, based on the **assumption** that the bulb is selected at random from the day's output. $S = \{GP1, GP2, DP1, DP2\}$, where $GP1$ is the event that a good bulb made by $P1$ is selected, and so on.

    $$P(GP1) = P(GP2 = P(DP1) = P(DP2) = 1/4$$

    Mark determined the probability that a defective bulb is chosen to be $P(DP1) + P(DP2) = 1/2$. (1) Is Mark's model satisfactory? How so? (2) Is Mark's conclusion correct? How so?

    (b) Mark's brother Bob suggested a simpler model, again based on the **assumption** that the bulb is chosen at random from the day's output. $S = \{G, D\}$, where G is the event that a good bulb is selected and D is the event that a defective bulb is chosen. $P(G) = P(D) = 1/2$. Bob pointed out that he and Mark had reached the same conclusion, so that each confirms the correctness of the other. (1) Would you agree or disagree with Bob's comment? (2) Is Bob's model satisfactory? (3) Is Bob's conclusion correct? (4) When asked for the probability that a defective bulb made in $P1$ is selected in terms of his model, Bob gave 1/4 as the answer. Would you agree or disagree? How so?

(c)   Should Mark's probability function be modified? How so? If modification is called for, how would you carry it out?

(d)   Should Bob's probability function be modified? If modification is called for, how would you carry it out?

2.   The Case of Jason's Well-Balanced Coin

Jason took a well-balanced coin from his pocket and asked his friend Andrew to help him determine the probability of throwing one head and one tail on two successive tosses of the coin. Andrew took $S = \{0,1,2\}$, where 0 is the event that no heads show in the two tosses, 1 is the event that one head shows in the two tosses, etc., as his sample space. He defined a probability function P by $P(0) = P(1) = P(2) = 1/3$, so that the probability of throwing one head and one tail is 1/3.

The relative frequency interpretation of this conclusion is that if a well-balanced coin is tossed twice in succession a large number of times, a head and tail will show approximately 33.3% of the time.

When Jason's well-balanced coin was tossed twice in succession 500 times (which involves 1000 tosses), a head and tail were observed to show 246 times.

(a)   Does this mean that Andrew's conclusion is not valid? How so?

(b)   Is Andrew's conclusion, interpreted in relative frequency terms, realistic? How so?

(c)   How is the discrepancy between the predicted relative frequency and the observed relative frequency to be explained?

(d) Set up your own probability model for tossing a coin, **assumed** to be well balanced, twice in succession.

(e) Should Andrew's probability function be modified? If modification is called for, how would you carry it out?

3.   Al Williams's Inspection Procedure

A lot of 20 items is known to contain 2 defectives. Consider an inspection procedure that consists of selecting 2 items at random from the shipment, one after the other, where the first item selected is not replaced before the second one is drawn. Al Williams was interested in finding the probability of the event B that the sample drawn contains 2 good items, and set up a probability model with sample space $S = \{G_1G_2, G_1D_2, D_1G_2, D_1D_2\}$, where G_1G_2 is the event that the first and second items drawn are good, G_1D_2 is the event that the first item drawn is good and the second item drawn is defective, and so on. Al assigned equal probabilities of ¼ to these sample points and found the probability to be ¼ that the sample drawn contains 2 good items.

(a) Is Al's conclusion valid?

(b) State the relative-frequency interpretation of Al's conclusion.

(c) Take 20 pennies, 2 that are new and shiny (to represent the 2 defective items in the lot) and 18 that have lost their luster (to represent the 18 good items in the lot), put them in a bag, shake the bag, and, without peeking, draw 2 pennies from the bag, one after the other. Repeat the process 300 times, record the occurrence of event B, and find the relative-frequency of B for the 300 repetitions of the process. Compare the result obtained with the relative-frequency interpretation of Al's conclusion.

(d) Does the result obtained affect the validity of Al's conclusion?

(e)   Is Al's probability model realistic?

(f)   How would you set up a probability model for the selection process?

(g)   Find the probability that the sample drawn contains 2 good items in your probability model, and interpret your result in relative-frequency terms.

(h)   Do the findings obtained in (c) support the results obtained in (g)?

4.   **Is There Something Fishy About Fred's Analysis?**

Fred Bass caught 9 fish, 3 of which were smaller than the law permits to be caught. A game warden inspects the catch by selecting 2 fish at random from Fred's bag and examining them. Some questions of interest to Fred are: what is the probability that no undersized fish are selected? What is the probability that at least one undersized fish is selected?

To answer these questions, he set up a probability model by taking as a sample space $S = \{f_1, f_2, f_3, f_4, f_5, f_6, f_7, f_8, f_9\}$, where f_1 is the event that fish 1 is selected, f_2 is the event that fish 2 is selected, and so on, and taking as his probability function P the one that assigns the same value, $\frac{1}{9}$, to each of the sample points. From this probability model Fred concluded that the probability that no undersized fish are selected is $C(6,2) / C(9,2) = \frac{5}{12}$, and that the probability that at least one undersized fish is selected is $1 - \frac{5}{12} = \frac{7}{12}$.

6.8 ANSWERS/DISCUSSION OF FOOD FOR THOUGHT

1.   **The Twolow Company**

(a)   (1) Mark's probability function is **not realistic** because it does not take into account the different proportions of good bulbs made by plant $P1$, good bulbs made by plant $P2$, etc. From the data given the proportion of good bulbs (of the total output of 8000 bulbs) made by $P1$ is 4950/8000, the proportion of defective bulbs made by $P1$ is 50/8000, etc.

(2)   Mark's conclusion is **valid with respect to his model,** but **not realistic** in terms of the process of choosing a bulb at random from the day's output with the given proportions of good and defective bulbs.

(b)   Bob's model is a open to the same sort of criticism leveled at his brother's model.

(c)   Yes, as follows: $P(GP1) = 4950/8000$, $P(GP2) = 2985/8000$, $P(DP1) = 50/8000$, $P(DP2) = 15/8000$.

(d)   Yes: $P(G) = 7935/8000$, $P(D) = 65/8000$.

2.   The Case of Jason's Well-Balanced Coin

(a)   No. This relative-frequency information has **no bearing on the validity of Andrew's conclusion.** The validity of Andrew's conclusion follows from the probability model that he set up.

(b)   No. The discrepancy between Andrew's valid conclusion interpreted in relative frequency terms (a head and tail will show approximately 33.3% of the time) and what actually happened (a head and tail showed 49.2% of the time) shows that Andrew's conclusion is **false about the behavior of well-balanced coins.**

(c)   Andrew's probability model is not a realistic model for describing the behavior of well-balanced coins.

(d) $S = \{HH, HT, TH, TT\}$, $P(HH) = \ldots = P(TT) = \frac{1}{4}$.

3.   **Al Williams's Inspection Procedure**

(a)   Al's conclusion is **valid** in that it follows as an inescapable consequence of his probability model.

$$P(B) = P(G_1 G_2) = \tfrac{1}{4}$$

(b)   If the sampling procedure is repeated a large number of times—that is, two items are drawn at random from a lot of twenty where the first item drawn is not replaced before the second is drawn, and this is done a large number of times—then the sample drawn will consist of two good items approximately 25% of the time.

(c)   In performing the underlying process 300 times the following was obtained:

$$R(B) = \frac{238}{300} = 0.793$$

Event B was found to occur 79.3% of the time, which differs markedly from the 25% value obtained from Al's probability model.

(d)   As established in part (a), Al's conclusion that the probability that the sample drawn consists of two good items is 0.25 is valid with respect to his model. **The data cited in part (c) has no bearing on the validity of his conclusion.**

(e)   The data cited in part (c) establishes that Al's conclusion is false about the underlying process when interpreted in relative-frequency terms. The appearance of a valid conclusion that differs strikingly from the results obtained by performing the process a large number of times establishes that Al's model is not a realistic one for the inspection procedure.

(f)   For convenience of discussion let us think of the items in the lot as labeled I_1, I_2,..., I_{20}. Another approach to the study of the sampling process is to take as our sample space S the events expressed by all combinations of 2 items that can be drawn from the lot of 20 items.

$$S = \{(I_1, I_2), (I_1, I_3),...,(I_{19}, I_{20})\}$$

The probability function P that best reflects the **assumed** randomness of the selection procedure is the one that assigns the same value, $\dfrac{1}{C(20,2)} = \dfrac{1}{190}$, to each sample point in S.

$$P(I_1, I_2) = ... = (I_{19}, I_{20}) = \frac{1}{190}$$

(g) As a **valid conclusion** of the probability model developed in (f) we obtain the following.

$$P(2 \text{ good items}) = \frac{C(18,2)}{C(20,2)} = 0.805$$

The relative-frequency interpretation of this conclusion is that if the sampling procedure is repeated a large number of times, the sample drawn will consist of two good items in the neighborhood of 80.5% of the time.

(h) The 79.3% value obtained by repeating the sampling procedure a large number of times is in close agreement with the 80.5% value obtained as a valid consequence of the probability model developed in (f). This provides **support for the realism of this model** for the sampling procedure.

4. **Is There Something Fishy About Fred's Analysis?**

In brief, the analysis is seriously flawed. Fred has an appropriate probability model for the process of catching one fish, but S is not a sample space for the process of catching two fish. The conclusion that the probability of catching two undersized fish is $C(6,2) / C(9,2) = \dfrac{5}{12}$ does not follow from his model for catching one fish.

Fred incorrectly formulated a probability model for the process to begin with, and then switched horses in midstream to obtain a conclusion that is valid with respect to some other probability model, but not the one he had set up. In short, a disaster.

The probability model with respect to which his conclusions are valid and that should have been set down is the following $S = \{(f_1, f_2), (f_1, f_3),...,(f_8, f_9)\}$, where (f_1, f_2) is the event that fish f_1 and f_2 are chosen, and so on. P is defined by $P(f_1, f_2) = P(f_1, f_3) = ... = P(f_8, f_9)$

$$= \frac{1}{C(9,2)} = \frac{1}{36}.$$

6.9 CASE 4: DEALING A POKER HAND OF 5 CARDS FROM A STANDARD DECK OF 52 CARDS

I return to this situation because of the student answer

$$\frac{C(4,3) \cdot C(48,2)}{C(52,5)}$$

(no it's-and-or-buts) that I usually receive in answer to the question, what is the probability of dealing a poker hand of 5 cards containing 3 kings (Sec. 4.2)?

It is important to address such a situation because of the role that combinatorics plays in teaching probability.

In the hope of disloging the entrenched It's-Just-a-Word-Problem thinking in probability I introduce to my students two probability models based on different **assumptions** that yield different valid conclusions.

But well-entrenched nonsense never yields easily, especially in probability where what is regarded by many as sacred combinatorics is involved.

Model 1

Looking ahead to reflecting by P an unbiased dealing of the hand and keeping in mind that our interest in a hand is in terms of its composition as opposed to the order in which the cards are dealt or arranged, we take as our sample space S the collection of all unordered hands of 5 cards that can be dealt from 52 cards.

$$S = \{(C_1, C_2, C_3, C_4, C_5), (C_1, C_2, C_3, C_4, C_6), ..., (C_{48}, C_{49}, C_{50}, C_{51}, C_{52})\}$$

The sample point $(C_1, C_2, C_3, C_4, C_5)$, for example, is the event that the hand dealt consists of cards C_1, C_2, C_3, C_4 and C_5, dealt in some order and arranged as one desires. The number of sample points in S is equal to the number of ways of choosing 5 of 52 cards without regard to order.

The **assumption** that the hand will be dealt in an unbiased manner (from a well-shuffled deck) is best reflected by the probability function P which assigns the same value, $\dfrac{1}{C(52,5)}$ to all $C(52, 5)$ sample points in S. We have:

$$P(C_1, C_2, C_3, C_4, C_5) = ... = P(C_{48}, C_{49}, C_{50}, C_{51}, C_{52}) = \frac{1}{C(52,5)}$$

With our probability model in place, we are ready to address the question posed. Since this is an equally-likely outcome model, probability questions reduce to counting questions.

We have:

$$P(3 \text{ kings}) = \frac{\text{Nu. of hands of 5 with 3 kings}}{\text{Nu. of sample points}}$$

$$= \frac{C(4,3) \cdot C(48,2)}{C(52,5)}$$

$$= 0.0017$$

Model 2

Let us suppose that it had been "arranged" with the dealer that he deal the poker player a hand consisting of three kings.

Clearly Model 1 is **not realistic** for this situation.

Assumption

The three kings are dealt at random from the four and the two non-kings are dealt at random from the remaining non-kings.

In formulating Model 2 we shall stay with the same sample space as the one employed for Model 1, but modify the probability function.

There are $C(4,3) = 4$ ways of choosing 3 of 4 kings and $C(48,2) = 1128$ ways of choosing 2 of 48 non-kings. Therefore there are $C(4,3) \cdot C(48,2) = 4512$ ways of choosing 3 of 4 kings and 2 of 48 non-kings.

The modified probability function, call it P_M, which realistically reflects the afore circumstance and **assumption** assigns 0 to the hands not consisting of 3 kings and 2 non-kings and

$$\frac{1}{C(4,3) \cdot C(48,2)} = \frac{1}{4512}$$

to the hands consisting of 3 kings and 2 non-kings.

The probability that the hand dealt contains 3 kings is, of course, 1, which in terms of Model 2, is the sum of the probabilities of the 4512 hands containing the 3 kings.

6.10 FOOD FOR THOUGHT

1. **Mark Ellis's Answer**

In answer to the exam question what is the probability that a hand of seven cards dealt from a standard deck of 52 cards has 3 queens, Mark Ellis replied:

$$\frac{C(4,3) \cdot C(48,4)}{C(52,7)}$$

The answer to this question is worth 10 points. If you were grading his paper would you give his answer full credit?

2.   Secretary Wilson's Wisdom?

Arnold Wilson, newly appointed Secretary of the recently established Department of Mathematical Affairs, was called on to make a few remarks at his installation as Department Secretary.

"One thing that separates mathematics from politics is that mathematics is definite, no ten sides to every story. Consider, for example, the event that there are 3 kings in a hand of 5 cards dealt from a standard deck of 52 cards. Its probability is

$$\frac{C(4,3) \cdot C(48,2)}{C(52,5)}$$

and that's a fact, no debate. If I submitted this statement in the form of a resolution before any committee or legislative body it would pass unanimously, no disagreement."

If you were on a committee or legislative body to which the following resolution was submitted, would you vote for it? If Yes, why? If No, why not?

Fact: The probability that a hand of 5 cards has 3 kings is $\dfrac{C(4,3) \cdot C(48,2)}{C(52,5)}$ period.

3.   A Poker Hand of 7 Cards

A poker hand of 7 cards is dealt from a standard deck of 52 cards.

Find the probability that the hand dealt contains:

(a)   3 aces,

(b)   3 aces and 2 queens

(c)   4 spades,

(d)   4 spades and 2 clubs,

Suppose that it had been "arranged" with the dealer that the hand to be dealt was to contain 2 aces.

(e)   How would you modify your probability function to take into account this situation?

(f)   What **assumption** underlies your models?

6.11 ANSWERS/DISCUSSION OF FOOD FOR THOUGHT

1.   **Mark Ellis's Answer**

It depends on the background information given. If the probability model with sample space

$$S = \{(C_1, C_2, C_3, C_4, C_5, C_6, C_7),...,(C_{46}, C_{47}, C_{48}, C_{49}, C_{50}, C_{51}, C_{52})\}$$

and probability function P which assigns 1/C(52,7) to each sample point in S were provided or if Mark Ellis stated this model as a preface to his reply, his reply would receive full credit.

If no such probability model preface were provided and Mark's reply stood alone in not-so-splendid isolation as an **absolute value in its own terms,** then his answer is fundamentally flawed. Such numerical answers do not stand alone in splendid isolation, but are valid consequences of a suitable probability model.

Misunderstanding of this basic point warrants a major credit deduction.

2.   **Secretary Wilson's Wisdom?**

No. The resolution is nonsense. Review Sec. 6.9, Case 4.

3.   **A Poker Hand of 7 Cards**

$S = \{(C_1, C_2, C_3, C_4, C_5, C_6, C_7),...,(C_{46}, C_{47}, C_{48}, C_{49}, C_{50}, C_{51}, C_{52})\}$; **assumption:** the deck is well-shuffled; the hand is dealt at random from the deck; P assigns $1/C(52, 7) = 1/133{,}784{,}560$ to each sample point in S.

(a) $\dfrac{C(4,3)\cdot C(48,4)}{C(52,7)}$ (b) $\dfrac{C(4,3)\cdot C(4,2)\cdot C(44,2)}{C(52,7)}$

(c) $\dfrac{C(13,4)\cdot C(39,3)}{C(52,7)}$ (d) $\dfrac{C(13,4)\cdot C(13,2)\cdot C(26,1)}{C(52,7)}$

(e) Take the same sample space S, modify the probability function P to P_M as follows:

$$P_M \text{ (any hand with 2 aces)} = \frac{1}{C(4,2)\cdot C(48,5)}$$

$$= \frac{1}{30{,}821{,}472}$$

P_M (any other hand) = 0

(f) The 2 aces are dealt at random from the 4 aces and the 5 other cards are dealt and random from the 48 non-aces

6.12 CASE 5: THE ONE MILE RACE

The Alumni Association of Ecap University has organized a one mile race to be run by 2 faculty, W.J. Adams and H. Lurier, and 3 alumni, J. Ross, M. Tilson and E. Kapp.

What is the probability that Adams finishes first?

Equally-Likely Outcome Model Let (ALRTK), for example denote the event Adams first, Lurier second, Ross third, Tilson fourth, Kapp fifth and take for S all permutations of the afore, which number 5! = 120. Take for P the probability function which assigns the same value to all sample points based on the **assumption** the the 120 finishes are equally-likely to occur.

$$P(\text{Adams first}) = \frac{24}{120} = \frac{1}{5}$$

Questions arise: Are the five runners in comparable physical condition, age, and running experience? No; Adams, for example, it has

been pointed out, tires quickly when the temperature is over 60 °F and the weather forecast is for 75 °F on the day the race is to be run.

> The equally-likely outcome model is legitimate in the probability model sense, but not in the sense of being realistic.

6.13 CASE 6: MARKOV FOR MARKETING?

I believe this is a striking example of a legitmate probability model, mathematically speaking, vs. real-world realism.

The late 1950s and 1960s saw the development of Markov chain models to describe consumer brand choice behavior. A starting point for many of these investigations is the view that a brand loyalty and brand switching matrix of probabilities can be constructed from data on sequences of consumer purchases.

$$T_1 = \begin{bmatrix} P_{11} & P_{12} & \cdots & P_{1n} \\ P_{21} & P_{22} & \cdots & P_{2n} \\ \cdot & \cdot & & \cdot \\ \cdot & \cdot & & \cdot \\ \cdot & \cdot & & \cdot \\ P_{n1} & P_{n2} & \cdots & P_{nn} \end{bmatrix}$$

The value p_{11}, for example, expresses the probability that the consumer, having bought brand 1 in the last period, will also purchase brand 1 in the next period. More generally, p_{ij} expresses the probability that the consumer, having bought brand i in the last period, will purchase brand j in the next period.

An early application of this sort is one undertaken by Benjamin Lipstein [7] concerning the test marketing of a new margarine, fictitiously called Electra, in the Chicago area from November 1958 to May 1959. In Lipstein's study the possible states a margarine buyer could be in were the following:

E_1: Electra Brand E_4: Aunt Mary's brand

E_2: Gloria E_5: Meadowlark brand

E_3: B-R Stores brand E_6: All other brands

E_7: Did not buy margarine during the time period

Lipstein's paper contains the brand loyalty and brand switching matrix of transition probabilities (Table 6.1) which represents the situation in the margarine market in Chicago shortly after the introduction of the new brand Electra.

Table 6.1

Next Period \ Original Period	Electra	Gloria	B-R	Aunt Mary's	Meadow-lark	Other	Did not buy
Electra	.12	.05	.03	.02	.04	.03	.05
Gloria	.05	.25	.02	.05	.01	.05	.03
B-R	.07	.03	.21	.01	.03	.03	.04
Aunt Mary's	.04	.02	.05	.23	.02	.04	.01
Meadowlark	.01	.02	.03	.04	.22	.05	.02
Other	.28	.26	.26	.25	.30	.23	.28
Did not buy	.41	.37	.40	.40	.38	.57	.57

Beware the Assumptions

But are Markov chain models realistic for the study of consumer brand choice behavior? A critical appraisal was given by A.S.C. Ehrenberg [1], who expressed the view that "**frequent public reference to Markov-brand switching models had not been matched by an obvious array of published demonstration of their practical effectiveness.**" On the basis of a detailed discussion and analysis, Ehrenberg concluded that

> the failure of the Markov brand-switching model to live up to its earlier public reputation need not be surprising if seen as an example of misguided but perhaps understandable enthusiasm for forcing an attractively simple piece of college mathematics (stationary Markov theory) onto repeat-buying and brand-switching data while:

1. Omitting to ensure that the data are of a technically suitable form to be modeled by the model.

2. Omitting to examine the crucial assumption involved.

3. Omitting any self-critical appraisal of the various concepts and analytical steps in the approach.

4. Omitting to gather any generalized empirical knowledge of repeat-buying and brand-switching behavior as such.

William F. Massey and Donald G. Morrison [12] expressed agreement with many of Ehrenberg's arguments that the simple Markov chain does not fit all, or even many, real brand-switching situations, but felt that Ehrenberg had been too harsh in his judgment. They expressed the view that the basic Markovian approach is fruitful and should not be abandoned.

In a reply [2], Ehrenberg took issue with Massey and Morrison and again raised the question, "Can we not bury Markov for Marketing?"

Ehrenberg Triumphant

As to Lipstein's study, it follows from his transition matrix that over the long term Electra brand would end up with about 4% of the margarine market. However, six months or so after its introduction Electra succeeded in capturing about 12% of the market. Electra had been effective in building up the percentage of buyers who having purchased Electra in one period, remained loyal to it in the next period, so that the 0.12 value in the first row, first column of Table 6.1, would have to be revised to 0.23.

The need to change transition probabilities had not been taken into account in Lipstein's model.

References

[1]   A.S.C. Ehrenberg. "An Appraisal of Markov Brand-Switching Models," *Journal of Marketing Research,* vol. 2, no. 4 (Nov. 1964), pp. 347-362.

[2]   _____ "On Clarifying M and M," *Journal of Marketing Research*, vol. 5, no. 2 (May 1968), pp. 228-29.

[3]   Jean E. Draper and Larry H. Nolan. "A Markov Chain Analysis of Brand Preference." *Journal of Advertising Research*, vol. 4, no. 3 (September 1964), pp. 33-39.

[4]   Frank Harary and Benjamin Lipstein. "The Dynamics of Brand Loyalty: A Markovian Approach." *Operations Research*, vol. 10., no. 1 (January-February 1962), pp. 19-40.

[5]   Jerome D. Herniter and John F. Mag. "Customer Behavior as a Markov Process." *Operations Research*, vol. 9, no. 1 (January-February 1961), pp. 105-122.

[6]   Ronald A. Howard. "Stochastic Process Models of Consumer Behavior." *Journal of Advertising Research*, vol. 3, no. 3 (September 1963), pp. 35-42. Reprinted in *Marketing Models: Quantitative Applications*, edited by R. L. Day and L.J. Parsons, pp. 104-117. Scranton, Pa.: Intext Educational Publishers, 1971.

[7]   Benjamin Lipstein. "The Dynamics of Brand Loyalty and Brand Switching." *Proceedings of the Fifth Annual Conference of the Advertising Research Foundation* (1959), pp. 101-108.

[8]   _____ "Tests for Test Marketing," *Harvard Business Review*, vol. 76 (March-April, 1961), pp. 365-369.

[9]   _____ "A Mathematical Model of Consumer Behavior." *Journal of Marketing Research*, vol. 2, no. 3 (August 1965), pp. 259-265. Reprinted in *Marketing Models: Quantitative Applications*, edited by R.L. Day and L.J. Parsons, pp. 65-79.

[10]   P.A. Longton and B.T. Warner. "A Mathematical Model for Marketing." *Metra*, vol. 1 (September 1962), pp. 297-310.

[11]   Richard B. Maffei. "Brand Preferences and Simple Markov Processes." *Operations Research,* vol. 8, no. 2 (March-April 1960), pp. 210-218.

[12]   William Massey and Donald Morrison. "Comments on Ehrenberg's Appraisal of Brand-Switching Models." *Journal of Marketing Research,* vol. 5, no. 2 (May 1968), pp. 225-227.

[13]   George P.H. Styan and Harry Smith Jr. "Markov Chains Applied to Marketing." *Journal of Marketing Research,* vol. 1, no. 1 (February 1964), pp. 50-55.

6.14 CASE 7: SUSAN'S DILEMMA

Susan's dilemma, I submit, serves as a revealing vehicle for thinking about the meaning of "correct" concerning validity vs. realism.

Susan Reti was interested in determining the probability that an even sum shows for the process of tossing a pair of well-balanced dice of uniform construction (one red and one green). She asked two of her friends, Rachael and Laura, if they would help her set up probability models to determine the probability of this event. Both were glad to do so.

Rachael's Model: Rachael took as her sample space S_1 the set of events described by all ordered pairs of integers between 1 and 6, inclusive, as shown in Table 6.2. The first number in each ordered pair specifies the number which shows on the red die and the second number specifies the number which shows on the green die.

Table 6.2

$$S_1 = \begin{cases} (1,1) & (1,2) & (1,3) & (1,4) & (1,5) & (1,6) \\ (2,1) & (2,2) & (2,3) & (2,4) & (2,5) & (2,6) \\ (3,1) & (3,2) & (3,3) & (3,4) & (3,5) & (3,6) \\ (4,1) & (4,2) & (4,3) & (4,4) & (4,5) & (4,6) \\ (5,1) & (5,2) & (5,3) & (5,4) & (5,5) & (5,6) \\ (6,1) & (6,2) & (6,3) & (6,4) & (6,5) & (6,6) \end{cases}$$

The information that the dice are well-balanced is best reflected by the probability function P which assigns the same value, 1/36, to each sample point in S_1, Rachael thought.

The sample points with even sums are found in alternate diagnols of Table 6.2, beginning with the first, and are 18 in number. Thus, from Rachael's model it follows that the probability of E, that an even sum shows, is:

$$P(E) = \frac{18}{36} = 0.50$$

The relative frequency interpretation of this conclusion is that if a pair of well-balanced dice are tossed a large number of times, an even sum will show approximately 50% of the time.

Laura's Model: Laura proceeded to analyze Susan's problem in a different way. She took as her sample space

$$S_2 = \{2, 3, 4, 5, 6, 7, 8, 9, 10, 11, 12\},$$

where 2 is the event that the sum of the numbers showing is 2, 3 is the event that the sum of the numbers showing is 3, and so on.

The information that the dice are well-balanced led Laura to the probability function P which assigns the same value, 1/11, to the eleven sample points in S_2.

$$E = \{2, 4, 6, 8, 10, 12\}, \text{ and Laura obtained}$$

$$P(E) = \frac{6}{11} = 0.55$$

for the probability that an even sum shows.

The relative frequency interpretation of this result is that if a pair of well-balanced dice are tossed a large number of times, an even sum will show approximately 55% of the time.

Susan was more confused than ever and turned to her cousin Jack for help with her problem.

Jack's Model: Jack took as his sample space S_3 the collection of events shown in Table 6.3.

Table 6.3

{1,1}	{1,2}	{1,3}	{1,4}	{1,5}	{1,6}
	{2,2}	{2,3}	{2,4}	{2,5}	{2,6}
		{3,3}	{3,4}	{3,5}	{3,6}
			{4,4}	{4,5}	{4,6}
				{5,5}	{5,6}
					{6,6}

Here {1,1} is the event that both dice show 1, {1,2} is the event that one die shows 1 and the other shows 2, and so on. There are 21 sample points, Jack observed. Since the dice are well-balanced, Jack was led to take as his probability function P the one which assigns the same value, 1/21, to each sample point. There are 12 sample points with even sum and this led Jack to conclude that $P(E)$, the probability that an even sum shows, is 12/21 or 0.57.

The relative frequency interpretation of this result is that if a pair of well-balanced dice are tossed a large number of times, an even sum will show approximately 57% of the time.

Susan's Question

Does $P(E) = 0.50$ (Rachael's Model), 0.55 (Laura's Model), or 0.57 (Jack's Model)? If mathematics is such a precise subject, how could this happen? Which result should I accept, Susan pondered? See Sec. 6.15, Food for Thought.

6.15 Food for Thought

1. **Susan's Dilemma**

   (a) **Validity**

   Which conclusions, if any, are valid? How so?

   (b) **Realism**

   How can it be determined which model, if any, is realistic? How so?

   (c) **Laura's Assumption**

   Is Laura's **assumption** realistic? How so?

   (d) **Jack's Assumption**

   Is **Jack's assumption** realistic? How so?

   (e) Return to **Jack's Assumption**

   If your answer to question (d) is no, how should Jack's probability function be modified to make it realistic for well-balanced dice.

2. **Herman's Dice**

   For his birthday, Herman received a pair of dice that, he was told, were well-balanced. On the basis of the probability model that assigns the same value, $\frac{1}{36}$, to each of the sample points in the sample space S shown in Table 6.4.

Table 6.4

$$S_1 = \begin{cases} (1,1) & (1,2) & (1,3) & (1,4) & (1,5) & (1,6) \\ (2,1) & (2,2) & (2,3) & (2,4) & (2,5) & (2,6) \\ (3,1) & (3,2) & (3,3) & (3,4) & (3,5) & (3,6) \\ (4,1) & (4,2) & (4,3) & (4,4) & (4,5) & (4,6) \\ (5,1) & (5,2) & (5,3) & (5,4) & (5,5) & (5,6) \\ (6,1) & (6,2) & (6,3) & (6,4) & (6,5) & (6,6) \end{cases}$$

Herman determined the probability of an even sum showing to be 0.50. He expected that an even sum would show in the neighborhood of 500 times for 1000 tosses of his dice and made betting plans accordingly.

Herman participated in a friendly game one evening, and after 1000 tosses of his dice an even sum had showed 200 times and Herman was $600 poorer. Disappointed, confused, and angry, Herman raised the following questions.

(a) If mathematics is such a precise subject, how could this happen?

(b) Isn't my conclusion correct?

(c) What went wrong?

6.16 ANSWERS/DISCUSSION OF FOOD FOR THOUGHT

1.   **Susan's Dilemma**

(a) **Validity**

All three conclusions are correct in the sense of being valid conclusions of their respective probability models; from the point of view of validity there is no conflict since these conclusions arise from different sources.

(b) **Realism**

To settle the question of realism, well-balanced dice would have to be tossed a large number of times and a record kept of how often an even sum shows. When this is done, it is found that an even sum shows in the neighborhood of 50 percent of the time.

This evidence also shows that $P(E) = 0.55$ and $P(E) = 0.57$ in terms of the relative frequency interpretation of probability for well-balanced dice are not realistic.

(c) **Laura's Assumption**

Laura's assumption is not realistic.

It is unrealistic to **assume**, for example, that a sum of 2 is as likely to occur as a sum of 7 when well-balanced dice are tossed; a sum of 2 can only occur in one way—when (1,1) shows; a sum of 7 can occur in six ways—when (1,6), (6,1), (2,5), (5,2), (3,4) or (4,3) show.

(d) **Jack's Assumption**

Jack's assumption, that the sample points in his sample space are equally likely to occur, is not an accurate reflection of the fairness of the dice. Consider, for example, the sample points {1,1} and {1,2}. {1,1} can only occur in one way, both dice must fall with 1 showing. {1,2} can occur in two ways, the first die shows 1 and the second die shows 2, the first die shows 2 and the second die shows 1. Thus for fair dice it is reasonable to expect that over the long run the event {1,2} will occur approximately twice as often as the event {1,1}.

(e)   **Return to Jack's Assumption**

Let x denote the probability to be assigned to $\{1,1\}$. The sample points $\{2,2\}$, $\{3,3\}$, $\{4,4\}$, $\{5,5\}$, $\{6,6\}$ should all be assigned the same probability x since they all can only occur in one way. The remaining sample points, $\{1,2\}$, $\{1,3\}$, $\{2,3\}$, and so on, should all be assigned a probability $2x$ since each can occur in two ways. We have six sample points which will be assigned a probability value x, yielding a sum of $6x$, and fifteen sample points which will be assigned a probability value $2x$, yielding a sum of $30x$. The total sum of the probabilities of the sample points is thus $36x$. Since this total sum must be equal to 1 we have

$$36x = 1, \text{ which yields } x = \frac{1}{36}.$$

Thus we are led to define P on Jack's sample space by $P(\{1,1\})$ $= P(\{2,2\}) = P(\{4,4\}) = P(\{5,5\}) = P(\{6,6\}) = 1/36$, and the probability of each of the remaining sample points—$\{1,2\}$, $\{1,3\}$, $\{2,3\}$, and so on—is 2/36.

2.   **Herman's Dice**

(a)   **If Mathematics is Such a Precise Subject, How Could This Happen?**

Herman is Hasty Harry's cousin and concerning the nature of mathematics their thinking is similar.

Herman, your faith in the precision of mathematical reasoning is misplaced. Mathematical reasoning is precise in the sense that its conclusions are valid with respect to the **assumptions** made; valid conclusions are not necessarily realistic statements about the process in question.

(b) **Isn't My Conclusion Correct?**

Your conclusion is correct only in the sense of being valid with respect to your probability model and the assumptions that it reflects; if we accept your probability model as a point of departure, then we must accept your conclusion as following from it.

(c) **What Went Wrong?**

What went wrong is that the probability model that came with your dice was not a realistic reflection of the nature of your dice. Your probability model is a good fit to well-balanced dice, and you had a pair of "loaded" dice.

7

The Other Side of the Coin:
Equally-Likely Outcome Models, Conditionally

7.1 PREFACE

That the applicability of equally - likely outcome models is not unequivocal is one lesson of Chapter 6. The other side of the coin is that they are realistic for many of the cases considered, conditionally.

An interesting problem is concerned with population size estimation, for which an equally-likely outcome model is the right one, conditionally.

7.2   POPULATION ESTIMATION

How many fish are in your favorite lake? How many raccoons are in your neighborhood? How many animals of your favorite kind are in the game reserve or national park? More generally, how many "whatever" are in your region of interest?

One approach to problems of this sort is based on what is called the **capture-release-recapture method.** We illustrate it by considering a fish population estimation problem, but the approach is applicable to the other situations noted as well.

We begin by catching a certain number of fish from the lake—100, say. These fish are tagged so as to be identifiable if caught again and are thrown back into the lake. We wait for a reasonable time to elapse to allow the fish to disperse (maybe a few days) and then catch another batch of fish, 200, say, and make note of how many in this batch were caught before. Let us suppose that one fish was twice caught.

Let N denote the number of fish in the lake. Our problem is to estimate N. To do this we set up a probability model for catching the second batch of 200 fish. As our sample space we take the events represented by the collection of all batches (combinations) of 200 fish that can be selected from N. There are $C(N, 200)$ such batches.

Based on the assumption that all fish in the lake are equally-likely to be caught, we take as our probability function P the one that assigns the same value, $\frac{1}{C(N,200)}$, to each sample point. We next determine the probability of catching 1 marked fish.

This probability is equal to:

$$\frac{\text{number of ways of catching 1 marked fish}}{C(N,\ 200)}$$

The number of ways of catching 1 marked fish in a batch of 200 fish is equal to the number of ways of catching 1 of 100 marked fish, $C(100, 1)$, times the number of ways of catching 199 of $N - 100$ unmarked fish, $C(N - 100, 199)$, which yields the product $C(100, 1) \cdot C(N - 100, 199)$. Thus:

$$P(1 \text{ marked fish is caught}) = \frac{C(100,1) \cdot C(N-100,199)}{C(N,200)} \qquad (7.1)$$

The right side of (7.1) depends on N. It varies as different numbers are substituted for N. Of special interest is that number which when substituted for N makes (7.1) assume its maximum value. This value is called the **maximum likelihood estimate of N**; that is, the maximum likelihood estimate of N is that number which maximizes the probability of catching the number of marked fish that were actually caught in the second batch.

We shall now show that the maximum likelihood estimate of N is 20,000.

The right side of (7.1) is a function of N, which we shall denote by $P(N)$.

$$P(N) = \frac{C(100,1) \cdot C(N-100,199)}{C(N,200)} \qquad (7.2)$$

We seek a positive integer value of N such that

$$P(N-1) \le P(N) \text{ and } P(N) \ge P(N+1) \qquad (7.3)$$

or equivalently:

$$\frac{P(N-1)}{P(N)} \le 1 \quad \text{and} \quad \frac{P(N)}{P(N+1)} > 1 \qquad (7.3)$$

Our first task is to determine and simplify $P(N-1)$, $P(N)$ and $P(N+1)$. From (7.2) we obtain

$$P(N) = \frac{\dfrac{100(N-100)\ldots(N-298)}{199\cdot198\ldots1}}{\dfrac{N(N-1)\ldots(N-199)}{200\cdot199\ldots1}}$$

$$= \frac{100(N-100)\ldots(N-298)}{199\cdot198\ldots1} \cdot \frac{200\cdot199\ldots1}{N(N-1)\ldots(N-199)}$$

$$= \frac{100(200)(N-100)\ldots(N-298)}{N(N-1)\ldots(N-199)} \tag{7.4}$$

For $P(N-1)$ we have

$$P(N-1) = \frac{C(100,1)\cdot C(N-1-100,199)}{C(N-1,200)}$$

$$= \frac{\dfrac{100(N-101)\ldots(N-299)}{199\cdot198\ldots1}}{\dfrac{(N-1)\ldots(N-200)}{200\cdot199\ldots1}}$$

Inverting and simplifying yields

$$P(N-1) = \frac{100(200)(N-101)\ldots(N-299)}{(N-1)\ldots(N-200)} \tag{7.5}$$

For $P(N+1)$ we have

$$P(N+1) = \frac{C(100,1)\cdot C(N+1-100,199)}{C(N+1,200)}$$

$$= \frac{\dfrac{100(N-99)\ldots(N-297)}{199\cdot198\ldots1}}{\dfrac{(N+1)\ldots(N-198)}{200\cdot199\ldots1}}$$

Inverting and simplifying yields

$$P(N+1) = \frac{100(200)(N-99)\ldots(N-297)}{(N+1)\ldots(N-198)} \tag{7.6}$$

From (7.4) and (7.5) we have

$$\frac{P(N-1)}{P(N)} = \frac{\dfrac{100(200)(N-101)\ldots(N-299)}{(N-1)\ldots(N-200)}}{\dfrac{100(200)(N-100)\ldots(N)-298}{(N-1)\ldots(N-199)}}$$

Inverting and canceling like terms yields

$$\frac{P(N-1)}{P(N)} = \frac{100(200)(N-101)\ldots(N-299)}{(N-1)\ldots(N-200)} \cdot \frac{N(N-1)\ldots(N-199)}{100(200)(N-100)\ldots(N-298)}$$

$$\frac{P(N-1)}{P(N)} = \frac{(N-299)N}{(N-200)(N-100)} \tag{7.7}$$

From (7.4) and (7.6) we have

$$\frac{P(N)}{P(N+1)} = \frac{\dfrac{100(200)(N-100)\ldots(N-298)}{N(N-1)\ldots(N-199)}}{\dfrac{100(200)(N-99)\ldots(N)-297}{(N+1)\ldots(N-198)}}$$

Inverting and canceling like terms yields

$$\frac{P(N)}{P(N+1)} = \frac{100(200)(N-100)\ldots(N-298)}{N(N-1)\ldots(N-199)} \cdot \frac{(N+1)\ldots(N-198)}{100(200)(N-99)\ldots(N-297)}$$

$$\frac{P(N)}{P(N+1)} = \frac{(N-298)(N+1)}{(N-199)(N-99)} \tag{7.8}$$

From (7.3), (7.7), and (7.8), our problem reduces to finding N such that:

$$\frac{(N-299)N}{(N-200)(N-100)} \leq 1 \quad \text{and} \quad \frac{(N-298)(N+1)}{(N-199)(N-99)} \geq 1$$

From the first of these conditions we obtain:

$$(N-299)\,N \le (N-200)\,(N-100)$$
$$N^2 - 299N \le N^2 - 300N + 20{,}000$$
$$N \le 20{,}000$$

From the second of these conditions we obtain:

$$(N-298)(N+1) \ge (N-199)(N-99)$$
$$N^2 - 297 - 2\,98 \ge N^2 - 298N + 19{,}701$$
$$N \ge 19{,}999$$

Thus:

$$19{,}999 \le N \le 20{,}000$$

We may take 19,999 or 20,000 as our maximum-likelihood estimate of the size of the fish population.

> **More generally,** the maximum-likelihood function that corresponds to catching k fish from the lake, tagging them and throwing them back, and catching a second batch of n fish that is observed to contain r tagged fish is defined by:

$$P(N) = \frac{C(k,r) \cdot C(N-k, n-r)}{C(N,n)}$$

In our example $k = 100$, $n = 200$, and $r = 1$.

By an analysis similar to the preceding one, it can be shown that the maximum-likelihood estimate of N is characterized by:

$$\frac{nk}{r} - 1 \le N \le \frac{nk}{r}$$

Black Bear Sightings

There have been more black bear sightings around the town of Charlotte and its residents have come to raise questions about the size of this population in the surrounding forest.

At the request of the town's council an applied statistics team at the local college undertook the project of obtaining a maximum likelihood estimate of the bear population. Six bears were caught, tagged, and released. Shortly thereafter 5 bears were caught and it was found that 1 had been previously caught.

In this situation k = 6, n = 5, and r = 1. Thus, the maximum-likelihood estimate of the bear population is:

$$N = \frac{nk}{r} = \frac{5(6)}{1} = 30$$

7.3   BEWARE THE ASSUMPTION

As precise as the maximum likelihood estimate of an animal population might impress us as being, we must not allow ourselves to forget that is what has been established is its precision in a mathematical sense. Its precision in a reality sense hinges on the realism of the second sample of animals caught being a random sample of the population.

The **realism of this assumption** depends on how well dispersed the tagged animals are throughout the population. **If this dimension is open to question, so is the realism of the maximum likelihood estimate of the population's size.**

7.4   FOOD FOR THOUGHT

1.   **The Fish Population of Lake Mark**

For the purpose of obtaining a maximum-likelihood estimate of the fish population of Lake Mark, 300 fish were caught, tagged, and released into the lake. Shortly thereafter, 500 fish were caught from the lake, and it was found that 2 had been previously caught. Let N denote the size of the fish population.

(a)   Set up a probability model for catching the 500 fish.

(b)   With respect to the given conditions, what is meant by the maximum-likelihood estimate of the fish population?

(c)   Determine the maximum-likelihood function for this situation.

(d)   Find the maximum-likelihood estimate of the size of the fish population.

(e)   Is it possible that the estimate is much too low? How so?

7.5   ANSWERS/DISCUSSION OF FOOD FOR THOUGHT

1.   **The Fish Population of Lake Mark**

(a)   Sample space S is the collection of events expressed by all combinations of 500 fish that can be selected from N fish. There are $C(N, 500)$ sample points in S.

Basic Assumption: The 300 fish that were initially caught, tagged, and released into the lake, were well-dispensed in the lake before the second sample of 500 fish were caught. We define P to be the function that assigns the same value, $1/C(N, 500)$, to each sample point in S. (b) The number that maximizes the probability of catching 2 tagged fish in the batch of 500 that were caught.

(c) $\dfrac{C(300,2)\cdot C(N-300,498)}{C(N,500)}$,   (d) 75,000

(e)   Yes; **if our basic assumption is unrealistic**, the realism of the 75,000 estimate is open to question and it may be much too low or much too high.

7.6   RANDOM SAMPLING IN THEORY AND PRACTICE

Consider a population of size N, $Q = \{x_1, x_2,...x_N\}$, and let us suppose that an unordered sample of size n is to be chosen at random from this population. As we have noted, when we say that a sample is to be chosen at random we have in mind the idea that there is to be no bias, deliberate or inadvertent, which favors certain samples of size n being chosen over others. The sampling procedure is to be an equal opportunity procedure; no favoritism.

The probability assignment P which best reflects random sampling is the one which assigns the same value,

$$\frac{1}{C(N, n)}$$

to each of the C(N, n) unordered samples.

While simple to envision in theory, random sampling is not simple to achieve in practice.

I return to this important issue in Chapter 8 of Part 2.

PART 2

If the Hypothesis of a Theorem in an Application Setting is NOT Satisfied, What Then?

PREFACE

The mainstay of mathematical proof is, If A then B. In application settings A is not satisfied exactly, but in some sense the hypothesis is "close" to A. May we still conclude B?

Random sampling is one situation in which this issue arises.

8

Is Random Sampling in Practice as Simple as It Sounds in Theory?

8.1 PREFACE

NO. Please consider:

8.2 HUXLEY COLLEGE'S SCHOLARSHIP FUND

To raise money for Huxley College's scholarship fund the student Social Science Society sold tickets at the College's homecoming affair. The prize was a 32 inch television set. The tickets were placed in a bowl and mixed. About midway through the festivities Huxley's President, blindfolded reached into the bowl and drew the winning ticket. What everyone expected and **assumed** was that the drawing was fair in the sense that there was no bias in the drawing which favored some tickets being drawn over others—that is, that it was drawn at random.

On the face of it the procedure seems reasonable enough for the task at hand, but how close does it come to satisfying the requirements of a random drawing? The problem is with the physical stirring of the tickets to achieve a "thorough mix." Obtaining a "thorough mix" becomes more and more difficult to achieve as the number of tickets increases, and it is not clear whether early, middle, or late ticket buyers might be favored and to what extent. Still, for the task at hand the degree of randomness achieved might be random enough, except possibly for those individuals who are willing to go to war over a TV set or what they consider the "principle of the matter."

8.3 GOING TO WAR

When it comes to going to war based on the outcome of a random drawing, the stakes are raised considerably higher.

In 1969 the administration of the draft in the United States to determine the order in which men born in 1950 would be drafted was changed to a lottery system. Three hundred sixty six capsules were prepared (for a leap year), each containing a birthdate. Each month's capsules were put into a separate box. The boxes were emptied into a drum, first those for January, followed by those for February, and so on for the subsequent months. The drum was rotated a few times, the capsules were poured into a bowl, and on December 1, 1969 the drawing was made. Those with birthdays on the capsules drawn first would be drafted first, and so on. If your birthday fell among those drawn last, there was a good chance that you would not be drafted. The results of the drawing are given in Table 8.1, from which we see that the earlier months, January through June got the larger share of the last-to-be-drafted numbers and the later months, July through December, got the larger share of the first-to-be drafted numbers.

Table 8.1

Month	1-122 (First Drafted)	123-244 (Middle)	245-366 (Last Drafted)
January	9	12	10
February	7	12	10
March	5	10	16
April	8	8	14
May	9	7	15
June	11	7	12
July	12	7	12
August	13	7	11
September	10	15	5
October	9	15	7
November	12	12	6
December	17	10	4

The results suggest the possibility that the earlier months' capsules were concentrated at the bottom of the bowl, while those of the later months were concentrated at the top and were more accessible for picking. Formal hypothesis tests of randomness did not support the hypothesis that a random drawing had been carried out (see, for example, [2] and [3]).

This period was one of great turbulence in American history and slogans such as "Draft Beer, not Students" and "Hell No, We Won't Go," were a prominent part of the scene. Perceptions of an unfair draft lottery on top of what was considered by many to be an indefensible draft for an indefensible war added fuel to a raging fire. In response to criticism the Selective Service modified its number selection mechanism for the draft lottery conducted in 1970.

8.4   RANDOM NUMBER GENERATION

Another way of obtaining a random sample which avoids the difficulties of obtaining a sufficiently thorough mix of tickets or capsules is by use of a table of random numbers.

The idea behind random number generation is that the digits from 0 to 9 are selected by a process such that each digit selected is independent of any other digit selected and all digits have the same likelihood or chance of being selected. The process, usually based on a computer program, is set in motion and thousands of digits are generated and recorded in the order in which they are generated.

8.5   RANDOM NUMBERS VS. PSEUDO-RANDOM NUMBERS

The study of many complex phenomena requires the generation of large streams of random numbers. It came as quite a shock when three scientists showed that five of the most often used computer programs for generating random numbers induced errors in the study of the behavior of atoms in a magnetic crystal because the numbers produced were not random, despite the fact that they passed several statistical tests for randomness [1]. The deviations from randomness were subtle and, although the pseudo-random numbers produced were satisfactory for many purposes, they were not satisfactory for the problem at hand.

Is it possible that no machine based system can produce truly random numbers? John von Neumann, regarded as the father of the modern computer, thought that the answer is yes. In an observation made in 1951 von Neumann expressed the view that anyone who believed a computer could produce truly random numbers was living is a state of sin. It may be that the best we can hope to do is produce pseudo-random numbers which are satisfactory for the purpose at hand, and that the truly random number is a mathematical ideal which cannot be attained. The question that arises in an applied situation then is, how close to random is random enough?

8.6 Deviation in Practice from the Requirements of Theory

As we appreciate from our daily experience, deviations from a stated norm may have consequences ranging from negligible to catastrophic. If a "little more or less" of a delicate herb than specified is added in the preparation of a gourmet meal, for example, the effect would in all likelihood be inconsequential. If we are off the norm by more than a "little," a gourmet delicacy could easily be turned into a gourmet disaster. Medication that is life saving when taken as specified might become life threatening when the prescribed dose is exceeded.

Deviation from mathematical requirements may, in a similar vein, have negligible or profound consequences, depending on the situation.

The profound lesson to be gleaned from these considerations is that when conditions are stipulated, whether they be in connection with sampling or of a more general nature, it is our obligation to insure, as best we can, that they are met if the conclusions derived from theory are to be in close agreement with the results obtained from practice.

References

1.   M. Browne, "Coin-Tossing Computers Found to Show Subtle Bias," *The New York Times,* Jan. 12, 1993, p. c1.

2.   C. Hawkins, J. Weber, *Statistical Analysis: Applications to Business and Economics* (Harper and Row, 1980), 297-303.

3.   J. Rosenblatt, J. Filliben, "Randomization and the Draft Lottery," *Science,* 171 (1971), 306-308.

8.7   FOOD FOR THOUGHT

1.   A Marketing Survey

To carry out a marketing survey on consumer preferences for kitchen appliances Elias Marketing Research Associates placed two interviewers on the busiest street in town to interview passersby. Does the sample of opinions obtained qualify as a random sample? Explain.

2.   Ecap University's Graduation Celebration

Tickets were sold at Ecap University's graduation celebration to help raise funds for the University's new library. The tickets sold were placed in a bowl as soon as they were sold. At the end of the graduation festivities the University's Library Director, Harriet Warren, reached into the bowl and choose a ticket at random. The ticket holder was awarded a newly published edition of Charles Dickens's collected works. Kevin Reynolds, who was among the first to purchase a ticket, protested that the drawing procedure was biased and demanded that ticket purchasers be given a refund or that the drawing be held again. "Your claim is not justified Mr. Reynolds," replied the Dean of Student Affairs. "Ms. Warren was blindfolded and the ticket was chosen at random."

Who is right?

8.8 ANSWERS/DISCUSSION OF FOOD FOR THOUGHT

1. **A Marketing Survey**

   No. The make-up of the busiest street is not necessarily (and probability not) the same as the make-up of all streets in the city.

2. **Ecap University's Graduation Celebration**

   If the tickets were thoroughly mixed so that it is **realistic to assume** that each ticket in the bowl had the same chance of being selected, then the dean is right. However, if this is not the case, then Reynolds is correct.

PART 3

What's the Price Tag for
Unrealistic Assumptions and Models?

PREFACE

Part 3 is not in the league of Parts 1 and 2 as necessities for the applications of finite math, but it shines a bright light on the importance of these parts for realistic decision making.

Decision making might have serious consequences (the Price Tag) when **unrealistic assumptions and models** are acted on. This is illustrated by examples in Chapter 9.

9

Consider:

9.1 INDIVIDUALS

Hasty Harry (Sec. 6.3) and his cousin Herman (Sec. 6.15, Food for Thought, Example 6) lost $300 and $600, respectively, because they placed wagers based on unrealistic probability models for their respective die and dice.

9.2 A COMPANY, ITS EMPLOYEE'S AND THEIR FAMILIES

Bottom-line Bob implemented the unrealistic production scheduling model LP-2 and the Austin Company's venture into the digital tape player market had to be written off as a disaster (Sec. 2.4). A number of employees lost their jobs, which was costly to them and their families in terms of wages and benefits.

9.3 TV NETWORKS, ADVERTISING AGENCIES, THEIR CLIENTS, AND THE TV VIEWING PUBLIC

The life span of a television program is determined by the public's reaction to it, which is measured by TV ratings. These ratings, produced by the Nielsen Company, estimate the audience in terms of the percentage of those sets in use which are turned to each channel, called a share, or in terms of the percentage of the total possible audience, sets on or off, called a rating. Shares and ratings are further broken down according to the sex and age of viewers so that advertisers can better focus their advertising campaigns. These numbers determine the buying and selling of billions of dollars of television air time. They mean life or death to television programs. The half-hour comedy *Good & Evil*, which had promising ingredients in terms of writing, acting and production talent, had a short life after its premiere in the fall of 1991 because of low initial ratings. In March 1992 NBC announced that they were dropping two successful shows, *Matlock* and *In the Heat of the Night*, because the demographic numbers favored older viewers while the network wished to build around a more youthful audience.

Since 1986 the data which underlie the ratings have been collected by a device called a people-meter. The remote control part of a people-meter rests on top of the television set. When the set is turned on, the meter prompts viewers to enter their identification number. Information is provided on what channels are being beamed into the household and who is watching them. Nielsen puts its people-meter into 4000 households selected at random—that is, without bias—from the approximately 93 million homes in America with television.

The people-meter data gathering system produced lower ratings for the networks than had been expected and a serious question arose as to whether this was because of the increased or decreased accuracy of this system over the method it replaced. The networks commissioned a study of the Nielsen methodology and two years later this Committee on Nationwide Television Audience measurement (CONTAM) issued a nine-volume report that was highly critical of the Nielsen system. The report found evidence of button fatigue—that over time people did not push the buttons that would insure data accuracy as they did in the beginning. CONTAM was highly critical of Nielsen's sampling procedures for obtaining the 4,000 households that make up their sample; random sampling was envisioned in the methodology, but the **actual sampling deviated significantly from random sampling**. From this came ratings which were highly suspect.

Random sampling was called for in theory, but not delivered in practice. This, with the problem of obtaining reliable viewer data, yielded unreliable ratings.

Nielsen overcame the statistical sampling problem, but it was still plagued by the problem of getting "honest" data from viewers in the sample selected.

Its people-meter system for eliciting viewing data has been described as too mechanical and as not being user friendly.

9.4   RECIPIENTS OF PUBLIC OR CORPORATE PENSION PLANS AND THEIR FAMILIES

"Many of America's largest pension funds are sticking to expectations of fat returns on their investments even after a decade of paltry gains, which could leave U.S. retirement plans facing an even deeper funding hole and taxpayers on the hook for huge additional contributions. . . .

The concern is that the reluctance to plan for smaller gains will understate the scale of the potential time bomb facing America's government and corporate pension plans."

'It's unrealistic,' "John Bogle, founder of mutual fund giant Vanguard, says of the return assumptions in place at most pension plans.

Pension funds at companies in the S&P 500 faced a $260 billion shortfall at the end of 2009, according to Standard & Poor's. Estimates of the fund deficits faced by state and local governments range from $500 billion to $1 trillion."

Reference

1.   D. Reilly, "Pension Gaps Loom Larger: Funds stick to 'Unrealistic Return Assumptions Threatening Bigger Shortfalls," *The Wall Street Journal*, 9/18-19/10, A1.

9.5   THE STUDY OF DISCIPLINES, ECONOMICS IN PARTICULAR

An engineer, a chemist, and an economist are marooned on an island and they find a can of tuna, but have no way to open it. Says the engineer: "Let's find a rock and crush the can open." Suggests the chemist: "That's impractical: A better way is to find some chemicals and blast it open."

Quoth the economist: "You people are truly misguided. There is only one way: Let's **assume** we have a can opener . . . ."

Gene Epstein, columnist for *Barron's,* reprises this tale to make an important point about commonly made **economic assumptions.** Specifically, Epstein considers the article "Labour Allocation in a Cooperative Enterprise" by Amartya Sen, who a week earlier had been announced as the 1998 winner of the Nobel Prize for Economics. Epstein notes that Sen sets down **assumptions** which, he assures us, are "serious but not especially odd in this branch of economics." They include "well-behaved utility and production functions; automatic fulfillment of the second order conditions of welfare maximization and of equilibrium; no uncertainty; perfectly competitive markets; homogeneity of labor . . .—a whole set of can openers" whose realism, Epstein points out, is "truly dizzying for the colossal naivete they reveal." More generally, "Your typical economist," Epstein further observes, "**assumes** such absurdities as 'no uncertainty' in markets, even though uncertainty is any market's middle name, and 'well-behaved utility and production functions' (read: unchanging consumer tastes and production processes), even though such good behavior is the rare exception rather than the rule—and he **assumes** so for only one reason: Because that way, **he gets to use a lot of math**" [1].

Epstein, as most of us, is not opposed to using a lot of math provided that the math conditions are suitable fits to the economic situation.

In the essay "The Future of Economics," *The Economist* puts it cogently: If models are to reveal anything, they muset be simpler than reality: the challenge is to simplify usefully" [2]. Also see [3] and [4].

References

1.   G. Epstein, "Is it Really Reasonable to Assume that the Newest Nobelist Deserved the Prize?" *Barron's,* Oct. 19, 1998.

2.   *The Economist,* "The Future of Economies", March 4, 2000

3.   M. Weinstein, "Students Seek Some Reality Amid the Math of Economics," *The New York Times,* Sept. 18, 1999.

4.   J. Galbraith, "How the Economists Got It Wrong", *The American Prospect,* Feb. 14, 2000.

9.6   THE NATION AND US ALL

The Consumer Price Index

The Consumer Price Index (CPI) value is intended to measure the behavior of inflation. It reflects the average change in prices over time of a market basket of about 80,000 goods and services each month in seven major groups. The CPI takes into account changes in the price of such items as food, clothing, housing, energy, transportation, medical and dental services, medical drugs, and other goods and services that people require for day-to-day living. Different urban areas, housing units and retail establishments around the country are taken into account in compiling this index. Then price changes for the various categories in each region are weighted to take into account the relative amount spent by that particular locale. Finally, local data is combined to obtain an overall average.

The CPI value generated as a valid consequence of the **postulates** of the underlying CPI model is published in the third week of each month by the Bureau of Labor Statistics (BLS). The monthly increments are added to yield an annual CPI figure.

Ripples

Indexed Programs. These are programs for which automatic increases in benefits are triggered by increases in the CPI. Social Security received by nearly 45 million beneficiaries, is the best known of these programs. Others include railroad retirement, with about 800,000 beneficiaries; supplemental social security income, with about 6.5 million recipients; veterans' compensation; and federal military and civilian employee pensions, paid to about 4 million retirees. The official poverty line rises each year in accord with the behavior of the CPI, which affects about 26 million recipients of food stamps, 25 million in subsidized child nutrition programs, and 5 million with federal student grants.

Many workers have union collective bargaining contracts that provide for automatic wage increases when the CPI value rises by a specified amount. The CPI may also be used to adjust such things as alimony and child support payments, attorney's fees, worker compensation payments, apartment, home and office rentals, and welfare payments. In addition to the preceding, movement of the CPI influences Congress on setting the personal tax exemption level and thereby touches most of us in some way.

Taxes. To protect taxpayers from the effects of inflation, taxes are adjusted in a number of ways. This includes tax brackets, which determine tax rates on income; personal exemption and standard deduction levels; earned income tax credit; limit on itemized deductions; and pension contribution limits.

Economic Statistics. Real growth in the gross domestic product (GDP) and productivity (i.e. growth adjusted for inflation) depends on the CPI.

The smaller the increase in the CPI, the smaller will be the additional benefits paid to beneficiaries, the smaller will be the cost to the government, the greater will be the tax obligation of taxpayers and government tax revenue, and the larger will be the GDP and productivity figures.

Reference

J. Berry, "The CPI's Wide Reach", *The Washington Post National Weekly Edition*, Dec. 23, 1996 - Jan 5, 1997.

Perspective

The CPI value is based on **postulates**. There has been disagreement about what **postulates** are appropriate to yield the CPI value which is the most realistic measure of inflation.

Consideration of the wide reach of the CPI value shows how the Nation and us all are dependent on the **postulates** chosen.

INDEX